[This is the last of the seven books by Kent
that I have edited — i.e virtually rewritten.
His organization is always excellent and
his analyses imaginative and just, but
his prose is often unclear and always
dully repetitious of formulas, syntax,
and transitions. Occasionally, too, Charles
Scribner and I have found scientific errors,
as well as absurd lapses about history and
past states of knowledge and opinion.]

JB (August 1989)

Reading the
Mind of God

ALSO BY JAMES TREFIL

The Dark Side of the Universe

Meditations at Sunset

Meditations at 10,000 Feet

A Scientist at the Seashore

The Moment of Creation

The Unexpected Vista

Are We Alone?
(with Robert T. Rood)

From Atoms to Quarks

Living in Space

Reading the Mind of God

In Search of the Principle of Universality

by James Trefil

Illustrations by Judith Peatross

CHARLES SCRIBNER'S SONS

NEW YORK

Charles Scribner's Sons
Macmillan Publishing Company
866 Third Avenue, New York, NY 10022
Collier Macmillan Canada, Inc.

Library of Congress Cataloging-in-Publication Data
Trefil, James S., 1938–
Reading the mind of God: in search of the principle of
universality / by James Trefil; illustrations by Judith Peatross.
p.   cm.
Includes index.
ISBN 0-684-18796-5
1. Science—Philosophy.   2. Uniformity of nature.
3. Physics—Philosophy.   I. Title.
Q175.T72   1989
501—dc19   89-4207   CIP

Macmillan books are available at special discounts for bulk purchases
for sales promotions, premiums, fund-raising, or educational use.
For details, contact:

Special Sales Director
Macmillan Publishing Company
866 Third Avenue
New York, NY 10022

10 9 8 7 6 5 4 3 2 1

Printed in the United States of America

Contents

Reading the
Mind of God

Introduction

THIS BOOK IS about an idea, one of the most astonishing and least appreciated ideas in modern science. I call it the principle of universality. It says that the laws of nature we discover here and now in our laboratories are true everywhere in the universe and have been in force for all time.

Why do I call this astonishing? Look at the earth for a moment, not as the expanses of land and ocean we call our home, but through the eyes of the astronomer. It is an insignificant ball of rock circling a very ordinary star located about two-thirds of way way out in the spiral arms of a very ordinary galaxy. This galaxy is one of hundreds in a supercluster of galaxies, and even the supercluster is only one of many in a universe that stretches billions of light-years in every direction. And yet, on that insignificant ball of rock the descendants of hominids that came down from the trees in Africa a few million years ago are able to discover the laws that make the whole marvelous machine work. If that's not astonishing, I don't know what is!

For in point of fact, there is no particular reason why the laws of nature that apply on the earth should govern a quasar ten billion light-years away. There is no reason why the principles that explain why your car coasts downhill should explain why a star suddenly becomes a supernova in a distant galaxy. Yet they do. How we came to discover this fact, and the consequences that flow from it, are the subject of this book.

Why do I say the principle of universality is largely unknown? This conclusion comes out of personal experience. Several years ago, I began giving lectures to a wide range of audiences, from physics departments at universities to Rotary clubs, on the question of the existence of extraterrestrials. As part of the talk, I would mention in an offhand way that we know the principle to be true, and then go on to explain the consequences for extraterrestrials. If the audience was made up of university scientists, this caused scarcely a ripple. We all knew about the principle—indeed, it is part of our folklore. But when I was speaking to an audience of nonscientists, the reaction was very different.

Anyone who has spent time lecturing develops a sense of when he or she has lost the audience. It's hard to give a precise description of what happens, but you sense a kind of dead silence wafting up from the seats. You feel a blankness—a suspicion that if you were to shout, the sound would disappear out there without trace or echo. This is exactly the feeling I got when I made my offhand comment about universality. The principle was not second nature to laymen, and they seemed surprised that I would assume it was.

I quickly learned to adjust my lecture, adding a few words about universality and why it was so important to modern science. But the experience kept bothering me. After all, educated people are at least aware that something called DNA governs heredity and that scientists are engaged in studies of the fundamental constituents of matter. How, then, could they not be equally aware of universality? The conclusion I finally came to was this: The principle of universality is so important for science that it is never explicitly taught. We learn it almost by osmosis. It pervades our work, particularly in fields like astronomy, but is seldom explicitly stated. After all, "everybody knows" it's true. Anthropologists tell me that this is a common attitude about deep myths among members of the initiated in any clan.

2

Nonscientists, of course, have not had this sort of education, so the principle is not embedded in their mental processes.

The more I thought about it, the more I realized that even though I, like my colleagues, believed in the principle, I had only the vaguest notion of how our predecessors had come to it and what the limits of experimental verification were. One of the benefits of this book for me, then, is that it has given me a chance to look deeply into this question and to share the results with my readers. I hope that some of the excitement of new discovery will come across in the following pages.

For most of the history of the human race, the principle of universality was not only unknown, but would have been regarded as the wildest sort of fantasy, if not heresy. For example, from Aristotle on, conventional scientific wisdom held that the laws that kept the moon in orbit around the earth had nothing to do with the laws that governed the fall of an apple from its tree. It took the genius of Isaac Newton, wandering in an English orchard in the seventeenth century, to see the underlying unity between these two seemingly disparate processes. From his time onward, the limits of universality have been pushed outward into the universe and backward in time until today cosmologists routinely talk about what happened in the first microsecond of existence and a few audacious souls even dream of creating new man-made universes! We have indeed come a long way from Newton's stroll through his orchard.

The story of the principle of universality is not a dry sequence of historical events, but a series of exciting (and sometimes serendipitous) advances made by men and women who, like all working scientists, could see only the dim outlines of the great edifice they were building. This fact is reflected in my choice of which events to talk about. In the final analysis, I chose the ones I did because I find in them a resonance with my own life and work.

During my career I have been lucky enough to have worked closely with a number of first-rate research groups and to have been involved, in a peripheral way, in some important discoveries. I have been able to see how the best scientific minds operate, to see what actually happens when the creative juices

3

start to flow. It is this perspective that I have tried to bring to the study of the historical events that led up to our present notions about universality, to an assessment of where we stand now, and to an explanation of the increased role the principle is going to play as we push the frontiers of knowledge into regions where it is difficult for conventional scientific experiment, not to mention human imagination, to follow.

ONE

In an English Orchard

"Besides," he thought, "in Chicago no one was respected until he'd rubbed somebody out. It was time for Aristotle to get his."

—ROBERT PERSIG
Zen and the Art of Motorcycle Maintenance

ACCORDING TO ISAAC NEWTON, it happened like this: He was walking in an orchard one day when he saw an apple fall to the ground. At the same time, he noticed the moon in the sky. He wondered whether the force that made the apple fall might not extend all the way out to the moon's orbit—after all, it was known to extend at least as far as the tops of the highest mountains. In one of those blinding flashes of insight that occur all too seldom in the history of science, Newton realized that if the force did extend that far, it could conceivably account for the fact that the moon remains in its orbit around the earth. The notion that the fall of the apple and the orbit of the moon are connected represents one of the greatest insights ever achieved by the human mind, and its consequences reach far indeed. How did Newton come to this conclusion, and why does it represent such a break with the past?

To answer these questions we have to understand the role played by the motion of the moon in the evolution of our ideas about the universe, and to do this we have to understand the intellectual problem that was posed by the fact that the moon and planets move in orbits.

That they do move in orbits was suggested by early Greek astronomers, who were well acquainted with the recurring transits of the planets across the sky. They assumed the earth was stationary and that the planets moved around it on celestial spheres. The idea was that all the planets, as well as the moon and sun, were embedded in separate crystal spheres, and the turning of the spheres carried all these bodies across the sky.*

The point I'd like to stress isn't the detailed structure of the Greek models of the universe, however, but their insistence on having the planets move in circles. In this aspect of their thought we can see the beginning of a rift in the sciences that was not finally healed until Newton.

The Tyranny of the Circle

One of the greatest problems that had to be overcome in the development of modern science was the hold that the idea of the circle seems to have on the human mind. It is well known to instructors in beginning physics courses, for example, that one of their hardest tasks is convincing students that there is nothing special about an object moving around a circle at uniform speed. The problem is that there *seems* to be something special about this situation, and the Greeks elevated this vague intuition to the status of a cosmological principle. They argued that since circular motion is perfect and without flaw, no outside

---

* The spheres were thought of as real, material things. It followed that nothing could move between the orbits of the planets; if something did, it would have to pass through the spheres, which would be impossible. This is why Aristotle, in his *Meteorology*, took such pains to show that comets were not heavenly objects at all, but conflagrations of vapors in the upper atmosphere.

interference was needed to make the celestial spheres rotate at a constant velocity.

With this argument, the Greeks not only provided an explanation for the movement of the stars and planets across the sky, they effectively isolated astronomy from the rest of science. In their system, the motion of objects in the heavens (the domain of astronomy) and the motion of objects on the earth (the domain of physics) were governed by different laws. On the earth, it is obvious that things left to themselves do not travel forever in circles. Since it seemed obvious that objects in the sky did, the only conclusion the Greeks could draw is that different laws must be in operation in the two domains. For them, the laws that govern the behavior of things on their earth are definitely not the same as those that govern the stars.

The vicissitudes of the elegant Greek geometrical universe were many. Upon closer observation, it was discovered that the tracks of planets across the sky did not correspond to the majestic rotation of celestial spheres. The notion of circular motion was salvaged (at the cost of making the system much more complex) by postulating that planets were attached to spheres-rolling-within-spheres. Even in the sixteenth century, when Copernicus modeled a solar system with the earth in orbit and the sun at the center, the tyranny of the circle over men's minds was so great that he, like his predecessors fifteen hundred years before, placed his planets on rolling spheres. As a consequence, his system was scarcely less complex than that of the Greeks.*

The first break with circular motion came in 1609 with Johannes Kepler. Kepler was a German mathematician who analyzed detailed observations of the positions of the planets and showed that their orbits were not circles, or even circles-within-circles, but ellipses. And while it may be true that the circle exerts a strange force on the human mind, the same thing cannot be said about the ellipse, an orbital shape that has nothing to recommend it except that it matches the data.

If Kepler's work demolished the old fixation with circular

---

* Copernicus claimed that his system was simpler, but if you read his work closely, you find that he actually has a few more spheres-within-spheres in his system than did his Greek predecessor, Claudius Ptolemy.

7

motion in the heavens, it also introduced a major difficulty. If the planets are not attached to celestial spheres that rotate forever because that is their nature, why do the planets move at all? And why do they move in elliptically shaped orbits? As so often happens in science, answering one question led to another one, equally important and even more difficult.

During the sixteenth and seventeenth centuries, the best scientific minds attacked this problem with little success. Kepler talked of "virtue"—something like a cosmic broom—emanating from the sun and sweeping the planets around in their courses. The French philosopher and mathematician René Descartes was somewhat more successful with a theory in which he imagined the universe to be full of giant, whirlpoollike vortices. The sun was at the center of such a vortex, and material streaming into the sun swept the planets around their orbits. Similarly, some of the planets (those with moons) were themselves the centers of secondary vortices, and this is how Descartes explained the fact that our own moon stays in its orbit.

These kinds of theories about the working of the solar system suffered from a number of difficulties. On the technical side, they all assumed that in order to keep an object in orbit it was necessary to supply a force to push it *along* its path. The late Nobel laureate Richard Feynman characterized this as the idea that you have to have angels following the planets along, flapping their wings to move them.*

On a deeper level, "virtue" and "vortices" preserve the ancient division between earth and sky. Although I suppose it would be possible to extend either theory to explain the fall of an apple, I'm not aware of any serious attempt to make this connection, either by the original authors or their followers. This means that as far as Newton's predecessors were concerned, the laws that governed the workings of the solar system had no connection to the laws that operated in everyday life. The fall of the apple had no more connection to the orbit of the moon than the number five has to the color green. The two

---

* To be fair, I should add that he also characterized Newton's explanation of orbits (see page 11) as the realization that the angels flapped their wings to push each planet toward the sun, rather than along its orbit.

sets of laws operated in different regions of the universe and described events that are, at bottom, completely disjointed.

The Newtonian Unification

It is hard to say when this notion of separateness of the earth and the heavens began to break down. In his later years, Newton claimed that the incident of the apple took place in 1666, when Cambridge University was closed because of the plague and he was spending eighteen months in isolation on the family estates. His findings were published in final form in 1687, in his monmumental three-volume *Principia mathematica*. Somewhere between these two dates, then, the separation of earth and sky, which had ruled men's minds for a millenium and a half, finally disappeared.

Leaving arguments about exact dates and priorities for later, let's look at just how the apple and the moon are related. Galileo had studied the behavior of bodies falling near the surface of the earth and had shown that projectiles such as cannonballs, that have some lateral motion and are simultaneously acted on by gravity, will trace out a particular mathematical curve known as a parabola. The higher the velocity with which the projectile is launched, the farther it will go. Some sample paths for various launching velocities are shown on the left in figure 1.1. (To make the following presentation easier, we show the projectiles being launched from a high cliff.) The straight downward line, corresponding to a projectile launched with no lateral velocity, is what would correspond to an apple falling from a tree.

FIGURE 1.1 (left)

FIGURE 1.1 (right)

You can see the connection between the falling object and the moon by looking at the right-hand side of figure 1.1. As we increase the lateral launch velocity of the projectile, the final landing point will move farther and farther away from the base of the cliff. You could, for example, imagine increasing the charge in a cannon until the cannonball is able to go all the way around the world and return to its starting point, as shown on the curve labeled A. If we neglect the resistance of the air, then when the cannonball returns to the launch point it will have the same velocity it did when it started out, so it will start around the circuit once more. A moment's reflection should convince you that the cannonball will just keep going around forever.

But this is exactly what the moon does in its orbit! The point of this argument is that there is a smooth transition between something falling directly to the base of the cliff and a satellite in orbit. In other words, we can go from the apple (which is clearly governed by earthly physics) to a satellite like the moon (which ought to be governed by heavenly physics) *without ever making a transition from one set of laws to another*. The force that causes the apple to fall determines how the satellite will move. Furthermore, it is clear from the example that this force, which we call gravity, is the only force acting on the satellite. This, in essence, was Newton's great insight — there is only one force of gravity, and it acts on apple and moon alike. This is why we refer to his hypotheses as the Law of *Universal* Gravitation.*

What Newton Actually Did

While the example given above is very useful for understanding why gravity must be universal, this is not the calculation that Newton said was inspired by the falling apple. What he did was more complex, but it leads to the same conclusion.

---

* For completeness, the precise statement of the law is that between any two bodies in the universe an attractive force of gravity exists, and the magnitude of the force is GMm/r^2. In this expression, M and m are the masses of the bodies, r their separation, and G is known as Newton's constant.

His thoughts went like this: We know that if it is indeed gravity that keeps the planets in orbit, then the force must be such that it makes the planet's "year" longer the farther away from the sun it is. For example, the earth circles the sun in one year, Jupiter takes nearly twelve. From this fact, Newton was able to deduce the way in which the force of gravity decreased as one moves away from the object exerting that force.

Knowing this, and knowing the force the earth exerts on the apple, we can deduce the force that the earth exerts on the moon. Newton then posed the following question: How far will the moon "fall" under the influence of this force in a given length of time?

Look at the situation as shown in figure 1.2. The moon is moving along, and at a point like the one labeled A it has some velocity along the line of its orbit. It is, therefore, just like the cannonball being shot from a cliff in that it continues to move along in the direction in which it is moving, but starts to fall under the influence of gravity as well. What Newton discovered was that the amount the moon "falls" in this situation is just enough to keep it moving along an elliptically shaped path. This, of course, is what Kepler had said orbits should be. From this seemingly simple calculation came the first inkling that the principle of universality might give us a key to understanding the entire universe.

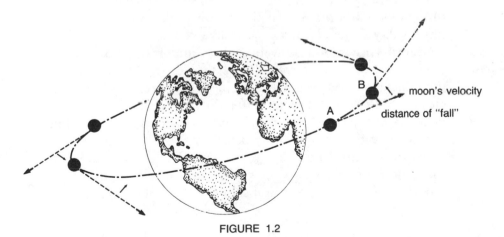

FIGURE 1.2

The Principle of Universality

It's hard to imagine an idea more at odds with the Greek tradition than Newton's notion that the apple and the moon are governed by the same force. What he told us is that there are no artificial distinctions in nature—no special laws that operate in one area but not in another. As far as gravity is concerned, the universe is a seamless web. The laws that operate here operate everywhere, without exception. It is this aspect of Newton's work that will concern us in this book, and that I will call the "principle of universality."

Implicit in Newton's universe is a degree of predictive capability unknown in previous cosmologies. All the bodies in the solar system obey the same law, a law that can be deduced by watching objects fall on the surface of the earth. We no longer have to guess at the rates of rotation of hypothetical celestial spheres, we can use a single law to tell us exactly how each planet must move. In fact, the metaphor most commonly used to describe Newton's universe is a clock. The planets, like hands, move along in stately fashion, while underneath the gears—the laws of nature—tick away and make the whole system work. You can think of God as having wound up the clock at the beginning, and what we're seeing now as the inevitable result of that initial setting in motion.

And although we have framed the discussion solely in terms of the moon and the planets, it is obvious that the principle of universality demands that objects like comets be governed by Universal Gravitation as well. This is important, because the appearance of comets was always thought to be erratic and unpredictable, and was never dealt with in previous cosmologies. The work of Edmond Halley, which we'll discuss in the next chapter, showed that the principle of universality could be extended even to those unruly denizens of the heavens.

Before we turn to that story, however, I would like to pause for a moment to discuss a few historical problems that cloud our view of the first great unification in the sciences.

Did Newton Really See the Apple?

The story of Newton's apple is such a wonderful, human way to describe one of the great intellectual advances in history that it has taken its place in our mythology, along with George Washington's cherry tree and Benjamin Franklin's kite. Consequently, I was very surprised to learn that there is some controversy among historians as to whether the incident really took place. Some reputable historians of science seem to take it at face value, others most emphatically do not.

The problem is that Newton was notorious for his vigorous (and sometimes unfair) championing of his claims for precedence in scientific discovery. The story of the apple has the effect of allowing him to claim that he discovered the law of gravitation in 1666 — twenty years before it actually appeared in print. This has led some scholars to suggest that Newton, either deliberately or through a quite normal fault in memory, embroidered a bit on the truth as to when, precisely, he had his flash of genius.

As far as I can see, the facts are these: everyone agrees that in his later years Newton recounted the story of the apple to his friends and claimed that he had calculated the orbit of the moon in 1666. The problem is that there are no documents to support this claim, and the first written records we have are from a much later period. Given this (rather typical) situation in history, what is one to do? On the one hand, it is reasonable to suppose that if Newton had, indeed, made a discovery of this magnitude he would have written *something* — a letter to a friend, a research note, a calculation scribbled on a scrap paper. On the other hand, documents do disappear, and the fact that a document doesn't exist today in no way implies that it never existed in the past. Scholars do the best they can with the situation and, as is often the case, manage to produce long and complex arguments supporting both sides of this particular issue.

There is, however, one body of evidence no one disputes, and that is the material contained in the *Principia*. In this book, Newton proceeds through a masterful exposition of his theory, cast in the form of a series of theorems and proofs modeled after Euclid's *Elements of Geometry*. Looking through this work, I can tell you one thing for certain: *Principia* is not the product of ideas that had occurred only recently to the author.

Throughout this book we will be looking at the process of scientific discovery. One aspect of this process that we will see over and over again is that when a new idea occurs to a scientist, it does not emerge in its final, polished form. It is often confused and muddled, and it may take many years, sometimes decades, before the consequences of the idea have been sufficiently thought out for an orderly presentation of the type found in the *Principia* to be put together.

My reading of the Newton-and-the-apple situation, then, is that the evidence we have in hand shows that Newton had the ideas well before they appeared in print. Whether this makes the story true or not is something I will leave to the historians.

Newton the Man

We in the twentieth century are in a somewhat unusual situation, because the best known and most brilliant of our scientists — Albert Einstein — was also a wonderful human being. I never cease to be amazed by the veneration in which he is still held, decades after his death. One example of this phenomenon can be seen in any university book store, where posters of Einstein often compete with Humphrey Bogart and Marilyn Monroe for the attention of the undergraduates. Because of Einstein, we have come to think of great scientists as mild, white haired, kindly old men — everybody's grandfather. But this attitude cannot be taken universally, especially not when discussing Isaac Newton.

By all accounts, Newton was not cast in the personal mold of Einstein. He was cold, arrogant, humorless, and something of a social climber. An assistant who worked with him for thirty years, for example, said he heard him laugh only once (when someone asked him whether reading Euclid's *Elements* was worth the time involved). In addition, in his later years Newton seems to have been obsessed with defending his priority in several key scientific discoveries, as well as in turning the entire edifice of British science into a personal fiefdom.

Perhaps two examples will make my point here. One of Newton's great obsessions was to establish his priority in discovering the mathematical form of the law of universal gravitation and the analysis of circular motion. His chief adversary here was Robert Hooke, best known today for his pioneering work in the study of solid materials. Hooke claimed to have been the first to come to the $1/r^2$ form of the law, basing his claims on his correspondence with Newton. Newton's response was immediate and violent:

> Now is this not very fine? Mathematicians that find out and settle and do all the business must content themselves with being nothing but dry calculators and drudges and another that does nothing but pretend and grasp at all things must carry away all the invention.

When some mutual friends tried to intercede and settle the squabble, Newton became so incensed that he went back and deleted all references to Hooke from the *Principia*, and even threatened not to publish the third volume. He maintained a special hatred for Hooke for the rest of his life, refusing to allow the Royal Society to publish his *Opticks* or to become the Society's president until after Hooke's death in 1703.

Was this reaction on Newton's part justified? It seems to me that an impartial reading of the work of these two men shows that Hooke did indeed come to understand the dynamics of circular motion before Newton, and that Newton eventually adopted Hooke's method of analysis. It is also true that Hooke's analysis of the motion of planets was flawed by his treatment of some of the technical points involved, and that Newton was the first to produce a theory that "had its pants on."

So what is one to make of this whole episode? To some extent, this is a matter of personal judgment. It might be useful to ask how such a conflict would be resolved today. Suppose the editor of a modern scientific journal were to be confronted with this priority fight and sent the papers to a typical scientist for advice. What would the reply likely be? If you accept my own reaction as typical of my colleagues, I would recommend that Newton be required to refer to Hooke's earlier work and then explain why his own was better. In fact, it was probably embarrassment over episodes like the Hooke–Newton controversy that led the scientific community to develop its present code of conduct.

Newton was also involved in a major dispute with the German philosopher and mathematician Gottfried Wilhelm Leibniz over who deserved credit for discovering the calculus. Leibniz complained to Newton, then president of the Royal Society, that he (Leibniz) had been accused of plagiarism by another member. Newton, in a display of arrogance that would have made a Chicago alderman blush, (1) packed a commision of inquiry with his friends, (2) wrote the committee report himself, and (3) wrote an anonymous review of the report that filled almost two volumes of the *Philosophical Transactions* (the Society's publication). He later remarked "pleasantly" to a colleague that he had "broke Leibniz's heart with his reply."

These sorts of examples could be multiplied endlessly, but I think you get the point. All in all, Newton was not the sort of man you'd want to invite to a cozy dinner party. But what does this tell us about the character of scientists? Which is the true paradigm, Newton or Einstein?

Whenever I consider this question, I start to think of all the scientists I've known and worked with. I think there is a kernel of truth in popular stereotypes. Researchers tend, on the average, to be more introspective and less socially oriented than the general population, although I've been to some parties that might cause you to question that judgement. When I think of the top flight of researchers — those who went on to win Nobel Prizes, for example — I can discern nothing that distinguishes them from their colleagues other than ability. There seems to be no correlation whatsoever between a person's status as a scientist and his or her behavior toward other people.

Let me recount two personal experiences to illustrate this point, both involving future Nobel laureates. When I was a new, wet-behind-the-ears postdoctoral fellow at MIT, I was scheduled to give a seminar. Its purpose was to introduce me to my new colleagues, and, needless to say, I was pretty nervous about it. Just before the talk was supposed to start, one of the senior people in the audience came up to me and said that he would have to leave the talk early to attend a meeting, and asked me not to worry when he left. This was a small act of kindness on his part, but one which I have always remembered—I hate to think of how I would have felt had he left in the middle of my talk without having warned me in advance. My appreciation grew in later years when I realized that this incident had taken place in the middle of a period when he was doing work that would revolutionize our concepts of the origins of the universe. That he would take time out from this work to consider the feelings of a new postdoc tells me that he is definitely on the Einstein side of the ledger.

On the other hand, I have come across some scientists who could probably out-Newton, Newton. One who comes to mind made a point of holding group meetings every evening at 8 P.M., seven days a week, so that his postdocs, most of them newly married with young children, wouldn't be tempted to fritter away their evenings with their families.* Even when he was traveling on another continent, he made sure that he called every office at 8 P.M.—you could stand in the hall and listen to the phones ring, one by one, down the line.

Experiences like these are what lead me to make the statement that ability and humanity are not correlated with each other. Apparently, I am not the only one to come to this conclusion. In recent times, the playwright Peter Schaeffer composed *Amadeus*, a fictionalized study of the conflict between Mozart and his less talented (but perhaps more human) contemporaries. One of the many messages that you can take away from this play is that in music, as in science, there is no such correlation.

Or, as I am fond of telling my students in less dramatic terms, there is no rule that says you can't be a jerk and a genius at the same time.

---

* I was, luckily, not one of these unfortunates.

TWO

Halley and His Comet

*Wherefore if [the comet] should return again about the
year 1758, candid posterity will not refuse to acknowledge that this was first discovered by an Englishman.*

—EDMOND HALLEY,
upon predicting the return
of the comet that now
bears his name

NO SCIENTIFIC THEORY, be it ever so elegant and beautiful, can be accepted until it has been tested. Newton's
theory of gravitation, with its attendant principle of
universality, is no exception to this rule. In the decades that
followed their publication, Newton's ideas were put to the test
with a vengeance. The great majority of these tests were the
drab, unexceptional sorts of things that are the lifeblood of
science — grubby, detailed calculations of orbits that were checked
by grubby, detailed observations. With one exception, the testing process involved nothing sexy, just plain solid scientific work.
That exception was the prediction made by the Oxford astronomer Edmond Halley that a particular comet that was bright
in the sky in 1682 would return seventy-six years later, in 1758.
The "recovery" of what we now call Halley's comet was one

of those spectacular events that serve to symbolize the success of new scientific ideas—in this case the ideas behind the Newtonian universe.

Edmond Halley

The lives of scientists are seldom as interesting as their ideas. The life of the mind just doesn't lend itself to treatment in swashbuckling movies. That's why Edmond Halley is such an interesting study. He was not, as many people imagine him to be, an academic recluse scanning the heavens from an ivory tower, but one of the most interesting figures of his time. He once faced a near mutiny at sea, he founded one of the first deep-sea diving companies, and, possibly, he acted as a secret agent for his government during his travels in Europe. In addition, he was a crucial figure in the maneuvering that led to the publication of the *Principia*.

Born in London in 1656 into the family of a prosperous soap manufacturer, Halley received the best education available in his time, ending up at Oxford University. He showed early promise as a scientist. Indeed, he published his first technical paper (on the mathematics of determining a planet's orbit) while still an undergraduate. Because of his father's wealth, he was in an enviable position—not only could he pursue his own interests without having to worry about making a living, he could afford to subsidize his own research efforts as well.

These advantages came into play in 1676 when, still an undergraduate, he conceived the idea of producing an astronomical map of the stars in the southern hemisphere. European astronomers had little information on southern skies, since they cannot be seen from northern latitudes. With a grant of three hundred pounds sterling per year from his father—a sizable income in those days—Halley set off for St. Helena in the southern Atlantic. This choice may seem odd, since the weather on that island tends to be moist and cloudy, but Halley's choice was governed by two factors: first, he believed from traveler's reports that the weather was usually clear; and second, the island was the southernmost English outpost at the time, which meant that he could set up shop without having to learn to operate in a foreign culture and language.

Despite the predictable problems with the climate (there is a description in his journals of how it was necessary to wipe the lenses of his telescopes off every fifteen minutes because of condensation) he succeeded in making the first coherent survey of the skies of the southern hemisphere. The results were published in 1679 and earned for the young man the nickname "The Southern Tycho."*

It is a characteristic of the scientific community that someone who provides the first reliable data in a new area of research becomes something of a celebrity. In our own time, this recognition takes the form of numerous invitations to give talks at universities and professional meetings — what I call an opportunity to hit the colloquium circuit. For Halley, with his private means, it took the form of an extended tour of Europe, complete with visits to all the major astronomical observatories. After his return to London he married and settled down to a quiet life of private astronomy. During this period he was part of the newly formed Royal Society, an association dedicated to the advancement of science. It was also during this period that he began working on the problem of explaining the tides, work that brought him into contact with Isaac Newton. This association was destined to have a major impact both on Halley and on the advancement of science, as we shall see presently.

But Halley's intellect was too wide ranging for him to be content with working in astronomy alone. The late seventeenth century was an era of exploration — an era when the new worlds that had been opened by the explorers were being charted and measured. It was also an era of invention and technical progress, and Halley, as a scientist, was probably as well situated as anyone to participate fully in these sorts of activities. Like most people, I had always associated him with the comet that bears his name, and I was amazed to find out how many contributions

* A reference to Tycho Brahe (1546 – 1601), the Danish astronomer whose observations of planetary orbits provided the basis for the work of Johannes Kepler.

he made to our understanding of the earth. I list a few just to demonstrate the man's versatility.

1. Halley established the relationship between height and atmospheric pressure—in effect, showing that the newly discovered laws that govern the behavior of gases explain the fact that the air is thinner on mountaintops than on the ground.*

2. He estimated the size of the atom by calculating the thickness of gold plate on a specimen of plated silver. He obtained an upper bound on the size of the atom by assuming the layer was one atom thick. His upper bound, in modern units, was about 0.000001 cm, which is roughly one hundred times larger than the actual size of the atom.

3. He put together detailed maps of winds and ocean currents worldwide. Based on ship's logs and his own later voyages, these maps were of obvious utility to a maritime nation like England.

In addition to these purely scientific ventures, Halley was one of the early developers of deep diving techniques. He apparently designed a diving bell into which air could be pumped and formed a company to engage in marine salvage operations. A publication called *Collection for Improvement of Husbandry and Trade*, a tip sheet for investors, commented that

> Some good Effect has been of Diving; but if Mr. Halley's should succeed, of which . . . he seems very sure; and, to my knowledge, has given such reasons for, . . . it would be very considerable.

The prices of shares in Halley's company were regularly quoted in the *Collection*, and he seems to have turned his study of atmospheric effects into a profitable venture.

Perhaps it was this business experience that led Isaac Newton, who was charged with overseeing a massive currency reform and reissue of silver coins, to appoint Halley as deputy comptroller of the mint at Chester, in Northern England. He took up his duties in 1697 and apparently discharged them well, despite being involved in a number of arguments (and a challenge to a duel with one of his colleagues).

---

* Halley showed that density and atmospheric pressure must fall off exponentially with altitude above sea level. We now know that this law is approximately true all the way to the top of the atmosphere.

At Sea

In fact, Halley was somewhat impatient with his term of office in Chester because he had an important commission from the Admiralty. To understand what it was about, you have to understand an important point about the common compass. We usually say that a compass needle "points north," and this is approximately true everywhere on the surface of the earth. When we look at the compass in the sort of detail needed for navigation at sea, however, we find that the needle does not point *exactly* north, but will deviate slightly from that direction. This so-called variation of the compass is caused partly by the fact that the north magnetic pole is in Greenland (instead of at the north pole proper), and partly by irregularities in the earth's magnetic field.

It was obviously extremely important both to the British navy and merchant seamen to have accurate measurements that would tell their sailors how to relate the readings on their compasses to true north, which can be defined as the direction of the North Star. Obviously, what had to be done was a series of measurements in which astronomical sightings were compared to compass readings so that the variation could be determined and charted. This was the task that Halley was assigned.

I had an interesting aspect of this problem pointed out to me by a friend, a civil engineer. "Can you imagine," he asked, "what it must have been like to be the first sailor to find out that the compass wasn't behaving the way the book says it's supposed to?" We don't, of course, know who that sailor was, but I'm willing to bet it happened on a cold spray-drenched night when the captain had left orders not to be disturbed.

In any case, Halley was given a ship called the *Paramour Pink** and set out in 1698 on a voyage to the south Atlantic and the West Indies. He soon had trouble with his crew, particularly with his first officer, a Lieutenant Harrison, who

---

* I wondered, too. It turns out that a "pink" is a type of vessel with three masts, a flat bottom, and a narrow stern. Your guess on "Paramour" is as good as mine.

was pleased grossly to affront me, as to tell me before my Officers and Seamen on Deck, and afterwards owned it under his own hand, that I was not only incapable to take charge of the Pink, but even of a longboat; upon which I desired him to keep to his Cabbin for that night . . . and accordingly I have watched in his steed ever since

Finding it impossible to operate the ship with officers like this, Halley returned to England, where the recalcitrant lieutenant resigned from the navy. Sailing again in 1699, Halley charted the compass deviations all the way to Antarctica, where his crew saw icebergs, then north to the Indies and Newfoundland, where an American fishing ship, mistaking them for pirates, opened fire and put "4 or 5 Shot through our rigging." Halley may be the only astronomer in history ever to have been shot at for being a pirate.

One interesting sidelight in this voyage occurred when the ship landed on the volcanic island of Trinidad, near Brazil.* His crew turned some goats and hogs loose to breed there, and as late as 1927 a visitor to the island reported that "wild goats and wild hogs liberated on the island in 1700 by the astronomer Halley roamed the ridges."

The measurements that Halley took on the voyage helped him complete his map of compass deviations, and his charts were published and widely used. His seafaring experience was seen in a different light a few years later, however, when the prestigious post of Savilian Professor at Oxford came open. Halley was in contention for the appointment, of course, but at least one observer commented that

Mr. Halley expects [the professorship], who now talks, swears, and drinks brandy like a sea-captain; so that I much fear his own ill behavior will deprive him of the advantage of this vacancy.

Well, it didn't. Halley was installed at Oxford in a small white house that still exists, nestled in the middle of that venerable university. He remained Savilian Professor for the rest of his life, even though he later moved to London to work with the

---

* Not to be confused with the larger island of the same name in the West Indies.

Royal Society and, in 1720, to be the Astronomer Royal. He continued his astronomical work until a few months before his death in 1742. It was while at Oxford, in fact, that he did the work that was to win him lasting fame.

But before going on to discuss the discovery of what is now known as Halley's comet, there is one more incident in Halley's life I'd like to relate. It has nothing whatsoever to do with the main theme of this book, but makes such a good story that I can't resist telling it. In 1698 Tsar Peter I of Russia (Peter the Great) visited England during his tour of Western Europe. During this tour the tsar often worked as a journeyman in shipyards, building ships with his own hands. While working in the London shipyards, he sent for Halley and apparently spent many evenings with him discussing the arts and sciences and "admitted him [Halley] familiarly to his table, and ranked him among the number of his friends."

The tsar stayed at a manor house in London with his entourage, but he seems to have had one quirk that no one had anticipated. He and his friends used to amuse themselves by placing one of their number in a wheelbarrow and them pushing him through the hedges. The visit wound up costing the tsar's host over three hundred pounds (a large sum of money in those days) to repair the damage to the hedges. Was Halley included in these pastimes? One biographer, a rather formal Englishman, says, "The legend that Tsar Peter trundled Halley through a hedge, though in character, lacks historical foundation."

Sure.

The Comet

The discovery of the recurrent appearances of what we now know as Halley's comet actually had its roots in events that happened long before Halley was appointed to his prestigious professorship. During the mid-1680s he and a number of prominent scientists, including Robert Hooke (see page 15), and the architect Christopher Wren, were thinking about how to calculate the motion of an object falling toward the earth or the sun under the influence of gravity.

It may seem strange to you that someone who knew the general form of the law of gravitation would not be able to apply that knowledge to a calculation of orbits, but the problem isn't as simple as it sounds, as you can see by consulting figure 2.1. At one point in the path of the object (look at the one labeled A), the force depends on the distance we have labeled r_A. In response to this force, the body will move to a new position such as the one labeled B. But at this point, the force depends on the distance r_B and will, in general, be different from the force at point A. Thus, a body falling toward the earth or the sun will experience a different force at each point along its path, and the shape of the path will, in turn, depend in a complex way on the body's response to this changing force. The forces experienced in the past determine where the body will be at any moment, and this, in turn, determines where the body will move (and the forces that will act on it) in the future.

Because of this complexity, one must use the mathematical techniques contained in the calculus to work out orbits, and the calculus had been newly invented by Newton and Leibniz. It is no wonder, then, that Halley decided to visit Newton at Cambridge in August 1684. A letter from a third person present gives an account of what happened at that meeting:

> After they had been some time together, the Dr. [Halley] asked him what he thought the curve would be that would be described by the planets. . . . Sir Isaac replied immediately that it would be an ellipsis. The Doctor, struck with joy and amazement, asked him how he knew it. Why, saith he, I have calculated it. Whereupon Dr. Halley asked him for the calculation without any further

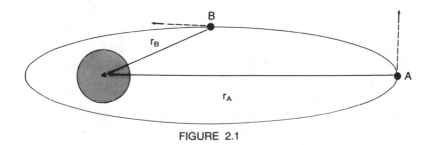

FIGURE 2.1

delay. Sir Isaac looked among his papers but could not find it, but he promised him to renew it and then to send it to him.

This theoretical development was important for a number of reasons. For one thing, it showed that the Law of Universal Gravitation did indeed predict orbits for the planets that were in keeping with Kepler's findings (see page 7). More importantly, it showed that *any* object bound to the sun, be it planet or comet, would have to trace out an elliptical orbit in the sky. This was crucial to Halley's thinking on the question of comets, because there was a great deal of controversy in his day about the way these bodies moved. At one stage in his career, for example, Halley believed that comets moved in straight lines, and that the apparent curvature of their paths was due to the motion of the earth in its orbit. It was the knowledge supplied by Newton's calculations that enabled Halley, years later, to predict the return of the comet that now bears his name.

Before describing this discovery, however, I would like to relate one other incident that grew out of these dealings with Newton. Halley was so taken with Newton's calculations that he undertook to help bring Newton's work to publication under the auspices of the Royal Society and served as editor for what eventually became the *Principia*. Unfortunately, when the manuscript was ready in 1686, the Royal Society had no money to pay for the printing, having exhausted its funds by bringing out *A History of Fishes*. Halley then reached into his own pocket to pay for the publication of the book, despite the fact that his own fortunes were then at a low ebb. His father had died in 1684, possibly a suicide, and he himself had recently taken a salaried position as clerk to the Royal Society to support his family. Without Halley's financial sacrifices, his painstaking editorial work, and his skill in handling Newton's difficult personality, the world might never have had the *Principia*.

In one sense, Halley was serving himself as well as posterity when he pushed for the publication of Newton's calculations on the mechanics of the solar system. Later in his life, when he was Savilian Professor at Oxford, he used the fact that the orbit of a comet must be an ellipse to study and catalogue some twenty-four comets which had been observed well enough to allow their orbits to be determined.* He found that three of these historical comets—those that appeared in 1531, 1607, and 1682—had orbits that were almost identical. They represented a comet that returned to the inner solar system about every seventy-six years.

Looking back through the historical records, Halley found other sightings that seemed to fit this pattern. In 1456, for example, a comet caused great fear in Europe. There is a story (probably not true) that the comet was excommunicated by Pope Calixtus III. In 1066, a bright comet appeared in the sky about the time William the Conqueror invaded England; its image can be seen in the Bayeux tapestry, a contemporary artwork depicting that event. Between 1066 and 1456 there are 390 (5 × 78) years, close enough to a repeating cycle to be suggestive.

Halley worried for some time about the fact that his comet didn't seem to come back at absolutely regular intervals, but eventually he realized that the gravitational effects of Jupiter and Saturn could serve to speed up or delay the return by a few years on any given cycle. Based on this work, he predicted that the comet would return in 1758.

Halley died sixteen years before his predicted return date, of course, but as 1758 approached, searching for the comet became something of an international sporting event. Detailed calculations predicted that the combined effects of Jupiter and Saturn would slow the comet by almost six hundred days, so that it wouldn't make its closest approach to the sun until mid 1759. Nonetheless, there was quite a competition to be the first to sight it on its way in from the outer reaches of the solar system. Much to the chagrin of the professionals, it was an amateur astronomer, Johann Georg Palitzsch, who first sighted the comet on Christmas Day 1758. Using a homemade eight foot telescope set up on his small farm near Dresden, he was the first to know that Halley's prediction had been fulfilled. The reappearance of the comet, on schedule, is referred to as the "recovery" of Halley's comet.

---

* It takes six numbers, or *elements* to determine the orbit of a comet. Two angles determine the plane of the orbit, two numbers determine the size and shape of the ellipse, two more the orientation of the ellipse in the plane and its relation to the sun. In his calculations, Halley simplified the calculation by approximating the ellipse by a parabola.

Since that time, Halley's comet has been a regular visitor in our skies. Its most recent visit, in 1986, was a bit disappointing. The comet was not visible from the earth when it was nearest the sun and hence most spectacular. Unfortunately for my younger readers, the next visit (in 2061) will be just as bad.

Knowing how to calculate the period of the comet has allowed astronomers to go back through the historical records and pick out earlier sightings of it. The best substantiated early observation was on May 25, 240 B.C., which was recorded by the Chinese. There are claims, somewhat more problematical, that the Chinese records contain evidence for a sighting in 1057 B.C., as well. All in all, if we include 1986 there are twenty-eight well-documented recurrences of Halley's comet in the historical record.

The Significance of the Recovery

The recovery of Halley's comet symbolized the success of the Newtonian clockwork universe. It's not that the scientific world regarded Newton's physics with skepticism until this new discovery forced acceptance — in fact, even before the publication of *Principia*, Newton was supported by a band of prominent partisans. The point is that Halley's prediction depended crucially on Newtonian physics, so when it was verified it provided a dramatic proof that the new ideas really did describe the universe.

The prediction was particularly dramatic because comets had always been regarded, even by astronomers, as objects whose appearances were essentially unpredictable. They simply didn't fit in any of the established models of the universe, and were classed with events such as earthquakes and volcanoes as erratic and unpredictable acts of God. Indeed, much of the folklore of comets, in which they are seen as omens of evil events, was based on the inability of scientists to predict when they would appear. By showing that the appearances of at least one comet could be understood in terms of the Law of Universal Gravitation, Halley cast the net of scientific rationalism over the last outpost of unpredictability in our solar system. In doing so, he extended the principle of universality from the planets alone to the entire solar system.

You may be wondering why Halley couldn't have looked at his sample and picked out those sightings separated by seventy-six years, without recourse to fancy calculations. There are two reasons why this approach wouldn't have worked. In the first place, there are many comet sightings in the record, and you could probably find a few sightings separated by almost any time interval you wanted, particularly if you allow (as Halley did) multiples of that basic interval to enter your data set. If you did this, however, you would have had no way of knowing whether the different sightings were of the same comet coming near the earth many times or of different comets that just happened to come by at the times they did by accident. The only way to resolve this question is to do as Halley did—show that the comets all have the same orbit and thus are identical.

The second reason is even more important. As we saw, the returns of Halley's comet are *not* exactly regular, but vary from one sighting to the next depending on the effects of the outer planets. Even if Halley had tried to pick out comets by looking at the dates of the sightings only, he would have missed his comet because, in point of fact, it doesn't come back every seventy-six years. Only a knowledge of the gravitational effects of the outer planets could have allowed him to see that comets with identical orbits but different arrival times could be one and the same.

It has become fashionable lately to deride the notion that science proceeds in an orderly way from observation to theory to prediction to verification. The claim seems to be that because science, like any human activity, depends on human vagaries and a measure of chance, there can be no such thing as a "scientific method" of the type outlined in textbooks. While there is some truth to this claim, I would also point out that the story of the recovery of Halley's comet is a beautiful illustration of the classical notion of scientific method. A theory was propounded, a prediction was made on the basis of that theory, and the prediction was verified by observation, lending weight and credibility to the theory. The story reminds us that stereo-types exist for a reason, and that even though there may be exceptions to general rules, those rules still constitute useful tools as we seek an understanding of the scientific process.

THREE

Science in Context

It was the zeitgeist, *out to put the* Thurn
und Taxis *ass in a sling.*

—THOMAS PYNCHON
The Crying of Lot 49

THE GERMAN WORD *zeitgeist*—the spirit of the time—
suggests an interesting notion. It implies that there are
certain ideas in the air at certain times, and that these
ideas can be seen manifested in every human endeavor—art,
music, literature, and even in science. Newton's clockwork universe, especially as it was worked out by a series of great mathematicians and physicists in the eighteenth century, is often cited
as an example of this phenomenon. It clearly influenced the
thinking of people in those times, and was itself (the argument
goes) influenced in turn.

At one level, this rather commonplace observation cannot be
faulted. Science always takes place in a cultural context, and it
would be foolish to say that it is ever wholly independent of
that context. A problem arises, however, when some thinkers
(most of whom are philosophers of science) take the argument

one step further. Since science is part and parcel of the ever-changing *zeitgeist*, the argument goes, there is no such thing as a scientific truth independent of cultural influences.

One way of dealing with this question is to ask another—a chicken-and-egg question.* Do scientific theories like Newton's come first, causing changes in cultural attitudes, or does the cultural matrix come first, so that the scientific theories are simply one more expression of the *zeitgeist*? In other words, do Newton's ideas express a truth about the external world that is independent of the Age of Reason, or are they nothing more than an expression of that age, equivalent in some sense to a symphony or poem? Obviously, if the latter statement is true, then a scientific theory cannot be objectively true, any more than a symphony can be.

You may think that I am overstating things here, setting up a straw man. It may seem incredible that anyone could seriously propose such an argument, but I have been involved in enough faculty club arguments to know that many academics—particularly philosophers and sociologists of science—do indeed hold these sorts of views. I have yet to meet a serious natural scientist who does, but as one progresses away from the hard sciences into the social sciences and the humanities, this sort of scientific relativism becomes more common, although I am happy to say that its proponents are far from being a majority. As it happens, the history of Newton and Halley makes a very good "laboratory" in which to examine the question of how scientific advances are influenced by the cultural context and whether the existence of this influence makes it impossible to define an objective scientific truth.

We can start by trying to get some sense of how different Newton's thought was from what preceded it and what followed. When I was a student at Oxford, I had the good fortune to be the lone scientist in a group of extravagantly seedy intellectuals. As we solved the world's problems over late-night ses-

---

* Incidentally, the answer to the original chicken-and-egg conundrum is "the egg." The reason I say this is simple: The first chicken must have evolved from an ancestor species through a mutation involving a change in the DNA structure. The egg laid by the ancestor would contain the mutated DNA, and would therefore eventually give rise to the first chicken.

sions with beer and cheese, I began to acquire an appreciation of medieval history — the specialty of several of my friends. One man was doing a thesis on the curriculum of the thirteenth-century university, a task that involved going through old books to see what they contained.* Occasionally he would translate a science or math text from the Latin and ask me to read it and comment. Thus I had the opportunity to take a guided tour of the way that some medieval scholars approached things that would be called scientific problems today.

This was a strange experience, something akin to what I imagine anthropologists must feel when they first encounter a new culture. I could usually understand what problem was being attacked, and, with some difficulty, piece together the arguments. In retrospect, I now realize that some of the strangeness arose from predictable sources — repeated theological comments, for example, or mathematical arguments more akin to the geometry of Euclid than to anything modern. In this respect, the medieval texts were not all that different from Newton's *Principia*, which has its share of theology and geometry as well. The real difference I saw in the old writings was the almost total lack of reference to observation or experiment. They followed the medieval Aristotelian ideal of using pure logic to arrive at conclusions — conclusions that were then held to be self-evident products of reason itself.† Alternatively, they sought to solve problems by appeals to the authority contained within accepted writings and texts.

One old story my friend told me illustrates this attitude toward observation as well as anything I could say. It seems that there was a debate concerning the number of teeth in a horse's mouth. One by one the scholars got up and cited their sources — one quoted Aristotle, another, one of the church fathers, and so on. Eventually a very junior member of the company rose and pointed out that there was a horse outside, and everyone could go out and count its teeth. At this suggestion,

* He was actually more interested in handwritten marginal notes than in the text, since such notes showed which books were actually used. I believe he was one of the first researchers to use infrared photography to bring out notes that had been erased.

† By using this method, medieval physical scientists ignored the fact that Aristotle was himself a keen observer of the natural world. He produced a very good survey of tidal life forms in the Aegean, for example.

according to the manuscript, the brothers "fell upon him, smote him hip and thigh, and cast him from the company of educated men."

This mode of thought, which holds that one learns about the world by reading what authorities had to say and then reasoning to conclusions, is so alien to us that it is hard to believe that people really used it. I know I had to fight the temptation to assume that the authors of the texts I was reading really knew better, but just wouldn't admit it. It took me a long time to realize that highly educated scholars — men every bit as intelligent as anyone around today — simply started from different premises than ours, and consequently went about their business differently. The science associated with that particular *zeitgeist* was not the same as ours. And if their method was not as successful as ours in explaining the natural world, finding such explanations was not a high priority task in the Middle Ages. Making the natural world a fit object for religious contemplation was the primary goal of scholarship, and making the theories fit observations was only tangentially related to that mission.

The story of how the intellectual climate in Europe changed between the end of the Middle Ages and the seventeenth century has been told too often and too well for me to repeat it here. For our purposes, we simply note that this change in outlook and the development of the new sciences occurred at roughly the same time: that the two were inextricably bound together.

In the same way, a somewhat less dramatic shift in outlook has taken place in this century. The theory of relativity and the development of quantum mechanics came on the scene at about the same time as profound changes were taking place in many other fields. Absolute values were out, relativism was in. Nonrepresentational art, atonality in music, and various "modern" and "postmodern" trends in literature have all been cited by different authors as examples of the new spirit of our age. Allan Bloom, in his book *The Closing of the American Mind* (Simon & Schuster, 1987), gives a detailed, scholarly discussion of this trend and its aftermath. Once again, a change in the kind of science that is being done seems to be associated with a change in the *zeitgeist*.

On the face of it, there is nothing exceptional in what I've been pointing out. The twentieth century is different from the eighteenth, which, in turn, is different from the thirteenth. Each has a different art, different literature, different ideas and goals, and there's no reason why they shouldn't have different science as well. But does this mean that we must deny the existence of a scientific truth that is independent of the men and women who discover it?

There is an important issue here, one that is intimately related to the major trend of our time. In many areas of human endeavor, the old confidence in traditional values has eroded, to be replaced by a kind of flabby relativism.* I use this term to denote a point of view that holds that there are no fixed values, no way of making choices between competing ideas, and hence no truth independent of social or political convention. This viewpoint is usually cloaked in rhetoric about the value of diversity and multicultural influences in American life. Indeed, acceptance of diversity has become the watchword in one part of our intellectual life after another. Many academics seem to believe that it is totally unfashionable to say that one point of view is better than others, so they have a hard time with the sciences, where judgments about the worth of ideas are made all the time. If novels can be deconstructed, they argue, why not scientific theories?

I have thought a lot about this question over the past few years. As a scientist, it is very important to me that the sciences be seen as an integral part of our culture, on an equal footing with the humanities. Even so, I have come to the conclusion that the sciences are different from other disciplines. At the risk of sounding terribly authoritarian and unfashionable to some, and of seeming to belabor the obvious to others, I would propose the following statement as a succinct summary of what I see that difference to be:

In science, there are right answers.

In order to explain what I mean by this sweeping pronouncement, I will have to give examples of fields where there are no

---

* It can't be repeated often enough that relativism has nothing in common with Einstein's theory of relativity except its name.

right answers so that we can contrast them with the sciences. I will choose to discuss philosophy in this context, primarily because it is the area of the humanities with which I am most familiar. The comments I make about philosophy, however, are equally applicable to a wide range of other subjects, so I hope my philospher friends won't feel they are being singled out for unfavorable attention.

As a student, I acquired a passion for philosophy, particularly the field of epistemology—the study of knowledge. I spent most of my spare time during my first few undergraduate years reading widely in this area. At Oxford, I was one of the first students to read for a special subject in the history and philosophy of science. At several stages in my education, I seriously considered dropping physics to study this subject. Yet, in the end, I found it profoundly unsatisfying. The problem I perceived was this: Philosophy does, indeed, address the fundamental questions— How is it possible to achieve knowledge? What is reality?— and so on. Because of this, it is extremely attractive. But in practice, I found that philosophical arguments are often the intellectual analogue of the old western practice of pulling the wagons into a circle to fight off the Indians. The idea is to find a position that is invulnerable to attack by other philosophers, and then stay in the safety of the wagons. The criterion for a successful philosophical argument, therefore, is that it is constructed in such a way that your colleagues and peers cannot disprove it.

It's not hard to see how such a situation could lead to the notion that all points of view are equally valid. Provided you are sufficiently clever, you can defend almost any philosophical proposition. There is no external test for the validity of your arguments; all you have to be is smarter than your opponents. How, then, can any one point of view be superior to any other?

And while I have made this point in terms of philosophy, it is equally true of most other fields in the humanities. I believe that it is a lifetime of work in the humanities that makes its practioners willing to believe that the same sort of relativism holds in the sciences. In the words of my friend and colleague

Don Hirsch, the problem is that they believe that "everything is English Lit."*

But haven't I admitted that there is a connection between science and the general culture in which it is embedded? If this is true, how can I claim some sort of universal truth for the products of scientific work? By admitting that there is a subjective element in science, haven't I opened the door to the same sort of flabby relativism I decry in the humanities?

To understand my resolution of this paradox, you have to make a distinction between two different parts of the scientific process: (1) producing a new idea or theory; (2) testing it against observation or experiment. I will argue that it is the first of these that is primarily bound by cultural restraints, while the second is largely free of them.

Let's take Newton's work as an example. Like all scientists before and after, he was confronted with a bewildering, varied complex of phenomena in the world—phenomena that display no apparent rhyme nor reason. His job: find the laws that provide the underlying order for them. There are no rules that tell a scientist how to carry out this task, no recipe for creativity. In confronting the essence of the universe, a scientist is in the same position as an artist. He or she has to reach inside and find a statement that covers the felt need, whether it is a poem unifying facts and feelings or a theory unifying facts and reason.

Much has been written about the act of creativity in science, and I have little to add to it. To me, scientific (or artistic) creativity remains mysterious. I don't know where ideas come from, even my own. The only point I want to make is that producing a scientific theory is not an end in itself, but the first step in a long process.

Once conceived, a scientific theory has to be logically consistent—making it so is the scientific equivalent of the "circling of the wagons" mentioned above. In this, the scientific theory is no different from a philosophical position or a treatise on literary criticism. It, too, has to withstand the attacks of the

---

* Hirsch's excellent book *Cultural Literacy* (Houghton Mifflin, 1987) gives powerful arguments against adopting this sort of attitude even in the humanities.

critics. But for the scientific theory, that's only the first step on the road to legitimacy. The crucial difference between the sciences and their sister disciplines lies in what happens after this first step has been taken.

Unlike the case in philosophy and the other humanities, the scientific theory has to meet a second test. Once a theory is formulated, it can be used to produce predictions about the physical world. These predictions, in turn, can be compared with objective reality. It makes no difference if your theory is logically consistent or if your colleagues aren't clever enough to find a fatal flaw in it. If it fails this second test, it is dead. Had Newton's law of gravitation failed to predict the return of Halley's comet (see chapter 2), it would have been forgotten long ago. Had it failed any of the huge number of tests to which it has been subjected since the seventeenth century, it would likewise have dropped from sight, to be replaced by some theory whose predictions were more correct.

This, then, is the fundamental difference between science and the humanities. In the former there is objective testing for correctness that is lacking in the latter. As I am fond of reminding my students, in science it is possible to start from reasonable premises, argue impeccably, and still make predictions that do not match the data. This is not the case in other disciplines. In the arts, questions of correctness are largely irrelevant, as they should be. In other fields, such as literary criticism and philosophy, they tend to be ignored when scholars circle their wagons.

So where does this leave us as regards the relation between science and culture? Apart from some trivial connections (for example, between technology and available instruments), it is clear that the process of testing predictions is largely independent of culture. Halley's comet either returns when predicted or it doesn't. So except for questions about what tests will be made and what will be construed to be sufficient evidence to adopt a theory, we can forget about cultural influences on the second step in the creation of a scientific theory.*

---

*This is not to say that such questions are unimportant or uninteresting. Indeed, I discuss them in some detail in my book *Meditations at 10,000 Feet* (Scribners, 1985). It's just that they are peripheral to the present discussion.

This is not true, however, of the actual creation of the theory. This is the act of an individual mind, and as such it is as subject to the constraints of cultural conditioning as any other human action. Newton could no more have conceived of Einstein's notions of relativity than he could have appreciated rock music. As a creature of the seventeenth century, it would have been inconceivable to him that there was not a "God's eye" frame of reference from which the universe had to be observed to be seen correctly. Einstein's argument that all observers see events differently yet find the same underlying laws of physics simply could not have entered Newton's mind. With our advantage of hindsight, we see that this restriction was an example of a culturally induced limitation. Newton was no more capable of transcending his *zeitgeist* than we are of transcending ours.

Thus, the kind of theories that can be proposed to explain the universe at any given time are, indeed, influenced by the cultural milieu in which the scientist finds himself. In this way, a scientific theory is no different from a novel or a painting. But once the theory is there, the process of verification proceeds in a way that is largely independent of culture, and only a theory that has survived this process can be accepted as a genuine law of nature.

There are several important points that follow from this analysis. In the first place, it tells us that there are no divisions in human creativity: the scientist and the poet may ponder different aspects of the universe, but they draw on the same wellsprings within the human psyche to do their work. This point is brought home to me forcefully every year when I teach a physics course for liberal arts students. My audience always includes a large number of aspiring writers and poets.* These students, who expected science to be as rigid and soulless as it is usually portrayed, are amazed to discover the same kind of questing, probing activities in physics that they find in their own fields. When they tell me, "This is just like poetry," I consider that my teaching has been successful, for I have shown them something of the full range of what goes into the activity we call science.

---

* In fact, one of my prize possessions, taped to the wall above my desk, is a term paper composed in the form of a poem.

A second important point that follows from this analysis is that it makes absolutely no difference where a theory comes from as far as the testing process is concerned. Once it is enunciated, it can be tested regardless of its source. This fact explains why so many apparently "crackpot" theories—relativity and continental drift being two examples—are eventually accepted by scientists. The internal integrity demanded by the testing process leads to the openmindedness that is a hallmark of the true scientist.

The decoupling of creation and verification in the scientific process is second nature to most scientists, and it gives rise to a somewhat unusual attitude on their part. My experience has been that most of my colleagues prefer not to think about the first phase of the process at all. The real scientific process, for them, starts once a theory has been written down and they can begin to test it against nature. Most of the formal training of a scientist, after all, is devoted to learning the skills needed to do this sort of testing. There is no formal training devoted to creativity, not even to the study of historical examples. This aspect of science is learned by example (if it is learned at all) during that period in graduate school when a student is apprenticed to a senior researcher and writes his or her Ph.D. thesis. In addition, most of a scientist's working life is devoted to developing and testing hypotheses, to carrying out the second step. Moments of creative insight are few and far between. No wonder that their attention is focused on the second part of the scientific process.

I think it is this emphasis on the process of objective verification that gives scientists their special mind-set and makes them so uncomfortable with the analyses of their work by sociologists and philosophers. These scholars tend to be more interested in the act of creation itself, or in the details of the way that information is passed around within a group of scientists—subjects that scientists themselves feel are peripheral to their work. It is this difference in focus that irritates most scientists when they read studies of science conducted by nonscientists. "Why are you spending so much time on these inessential things when the real heart of science is in the verification process?" would be a typical response.* Because they are looking at distinct and separate parts of the scientific process, the two groups just can't seem to agree on what they're seeing. I know that I often wonder, when reading descriptions of the scientific process by sociologists, if this is how an atom would feel if it could read a quantum mechanics textbook.

Perhaps I can best illustrate this point by describing a seminar I attended a few years ago. It was delivered by a young sociologist who was addressing the question of how the scientific community reacted to the theory that the dinosaurs had been driven to extinction by the impact of an extraterrestrial body. Her work was very good, and she traced out in detail all the infighting and personal conflict that normally accompany the introduction of new ideas. But as I listened, I began to feel that familiar sense of irritation. Amidst all her talk of disputes with editors of scientific journals and phone calls to reporters, there was no mention of the slow accumulation of supporting evidence that had finally driven me (and others) to accept the theory. In fact, acceptance or rejection seemed to her to depend only on the phone calls and maneuvering, not on anything else.

In the discussion that followed, this point was raised repeatedly by many of us, but we just couldn't get her to see that it was important. On our side, we simply could not convince ourselves that an article in *The New York Times* or *Science* could play any role in the verification process (except, of course, in bringing the theory to the attention of a large number of scientists). Finally, one senior paleontologist present expressed our frustration perfectly when he burst out with the question, "Is it really news to sociologists that evidence counts?"

In the end, then, this is what the debates about the social context of science come down to. Scientists focus their attention on the process of objective verification of theories and claim their results are largely culture-free. Others concentrate on the process by which theories are created and claim that because they are conceived in a given culture, scientific results have no

---

*There are, of course, other complaints based on things like oversimplification of complex interpersonal relationships and the uncanny ability of some social scientists to see what they want to see in any situation. But don't get me started on these.

validity outside of that culture. My opinion is that while neither view is completely correct, the latter results from a fundamental misunderstanding of the way that science works, and can safely be disregarded.

After all, gravity pulls on the Bushman as well as on the European.

FOUR

Reading the Mind of God

In enterprise of martial kind
When there was any fighting
He led his regiment from behind
He found it less exciting.

—W. S. GILBERT and
SIR ARTHUR SULLIVAN
The Gondoliers

WILLIAM HERSCHEL, WHOSE life's work was to revolutionize the science of astronomy, began his career a second oboe. That is, he followed his father and older brother, both musicians, into the regimental band of the Hanoverian Guards in his native Germany. He was then a teenager, and the German states were being drawn inexorably into the Seven Years War (known in this country as the French and Indian War). In 1755 the Hanoverian troops, along with their band, were sent to England because an invasion by the French was expected. The invasion never materialized, but it was during this visit that Herschel first saw the country that was to become his home for most of his life.

Returning to Germany in 1757, the regiment was involved in a disastrous defeat at the town of Hastenbock, about twenty-

43

five miles from Hanover. This was Herschel's first (and last) taste of battle. In later life, he wrote that

> we were so near the field of action as to be within the reach of gunshot; when this happened, my father advised me to look to my own safety. Accordingly, I left the engagement and took the road to Hanover.

Reading biographies of Herschel written in the nineteenth century, you find endless pages devoted to a debate on whether or not he was a deserter. The problem was that the rulers of Hanover assumed the throne of England soon after the war, and Herschel was a major recipient of royal favors throughout his life. Some biographers argued that George III had had to pardon Herschel for desertion so that he could enjoy this support. But this is just an example of imposing contemporary standards on the past. In fact, Herschel was not a soldier, but a musician. His discharge from the Guards (issued in 1762, years after his "desertion") has the word "soldier" crossed out and "oboist" written in by hand. In any case, I suspect that the officer corps had a lot of things on its mind in the wake of the defeat besides the behavior of a young second oboe who, in any case, returned to the regiment in a matter of days.

This episode apparently convinced Herschel that following his father's footsteps in a military band was not for him. Moving to England in 1759, he began a career as an itinerant musician, giving lessons here, leading bands and orchestras there. In 1766 he took the position of organist at a new chapel in the town of Bath. It was here that he began the work for which he is known today, and transformed himself from a run-of-the-mill professional musician to a major figure in astronomy.

It is extremely difficult to see anything in Herschel's early life to suggest that he would become a scientist. His education, while adequate for his time, was certainly not outstanding. In his diaries, he talks about looking at the stars with his father and long, late night philosophical discussions with his family and friends. But this is hardly enough to explain his sudden decision, five years after coming to Bath, to become an astronomer.

Herschel kept a terse diary throughout his life. Up to 1773, it is full of entries like "gave a concert at Chapel," or "began again to teach my resident scholars." It chronicles the life of a successful and progressively more prosperous musician on a round of composition, concerts, and teaching. Then, on April 19, 1773, we suddenly find the entry "Bought a quadrant* and Emerson's *Trigonometry*."

That's it: from that point on the diary records his acquisition of lenses, telescope tubes, and all the rest of the paraphernalia of the amateur astronomer. There is no explanation at all of this abrupt change. Some scholars speculate that he may have begun studying mathematics because of some connection to music, but that seems pretty thin to me. I prefer to leave Herschel's sudden change of career as one of those amiable mysteries whose solution we will probably never discover.

Whatever the reason for this sudden transition in his life, the results were plain. He began building his own instruments—telescopes larger and of better quality that those generally available at the time. With characteristic thoroughness, he turned the entire house in Bath into a telescope factory and observatory, so that at one point his sister Caroline wrote in her diary:

> to my sorrow I saw every room turned into a workshop. . . . Alex [a brother] put up a huge turning machine in a bedroom.

During this period of his life Herschel worked with an energy that inspires a sense of awe, envy, and disbelief in the modern reader. After giving lessons to private students throughout the day, he would spend the evening working on his telescopes, then repair to the backyard for a night out in the cold, surveying the stars. When did he sleep?

He did, of course, have one advantage over his modern counterparts. In a medium-sized town like Bath, the skies at night were almost as dark as they were in the countryside. There were no neon signs in those days, and no smog. Thus, Herschel had only to take his telescopes out into his backyard to test the results of his handiwork and carry out what he called a "review of the heavens"—a systematic survey of the stars, the planets, and the moon.

---

* This is an astronomical instrument used to ascertain the distance of a star above the horizon.

It was while in his backyard in Bath, using a homemade telescope, that Herschel made the discovery that was to take him forever from the world of music and start him on his life's work as an astronomer. On March 13, 1781, between ten and eleven o'clock, he noted an object that he described as "either a nebulous star or a comet." He recognized from the object's appearance that it wasn't an ordinary star—it was an extended bright spot, not a point. When he went back to check the sighting a few days later, the object had moved relative to the background stars. This ruled out the possibility that it was itself a star, and Herschel concluded that it must be a previously unknown comet.

The news of the new comet was sent out—first to astronomers in England, then to the rest of Europe. As measurements of the object's position accumulated, it became clear that it was not a comet. In fact, it was the planet we now call Uranus—the innermost planet that cannot be seen with the unaided eye.

The impact of Herschel's discovery on Europe was immediate and intense. After all, no human being had ever discovered a planet before. The effect of the discovery on Herschel's life was equally profound. Within six months he had presented his results to the Royal Society and been unanimously elected to membership. Within two years he received a pension of two hundred pounds a year from King George, enough to secure financial independence and allow him to devote himself full time to astronomy.

The election to the Royal Society meant that Herschel was now in contact with all the men of science in Great Britain. From an isolated amateur in a backyard in Bath, he was suddenly elevated to the status of equal with the best scientists of his day. It is clear from his correspondence that this exposure had an important effect on his work—his new friends polished up the rough edges and helped him to speak the language of science.

Viewed in the grand sweep of scientific progress, the discovery of Uranus was not really a major event. It did have some effect on the principle of universality, simply because Newton's laws of gravitation were found to describe its orbit as well as those of the visible planets. But when all is said and done, it may well be that the the most important consequence of the discovery was that it freed William Herschel to devote all of his time to astronomy.

Reading the Mind of God

Just as we don't know exactly what it was that led Herschel to give up a thriving career as a musician to embrace astronomy, we don't know exactly what it was that brought him to the systematic study of the skies that was to be his most lasting contribution to science. Writing in 1811, near the end of a long and productive life, he said, "A knowledge of the construction of the heavens has always been the ultimate object of my observations."

I suppose it really doesn't matter whether we know why Herschel adopted this particular goal, just as it doesn't matter whether we know why he became an astronomer. What I find fascinating in this story is the sheer audacity of the man. Here he was, a journeyman musician with a telescope in his backyard in a provincial English town, deciding to take on the task of unraveling the structure of the universe. Like many of the amateur scientists of his time, he lacked the education, the background, and the scientific connections to take on a major project like this. But, as many people did in those days, but he took it on anyway. Where did he—and they—get the nerve?

My wonder at his setting himself this task is compounded by the fact that up until Herschel's time, astronomers had fixed their attention almost exclusively on the solar system and had paid little attention to the stars. The stars appear as fixed points of light when you see them with your naked eye, and they appear as fixed points of light when you see them with a telescope, even the best telescopes around today. Beyond the fact that they existed, there seemed to be little of interest one could say about them.

But Herschel had a grand vision of what needed to be done —a vision that borrowed some of its main components from the new science of geology that was being developed in the latter half of the eighteenth century. If people could go out and learn about the structure and history of the earth by looking at rocks, why couldn't astronomers do the same by looking at stars? In a paper read to the Royal Society in 1784 and titled "Account of some Observations tending to investigate the Construction of the Heavens," Herschel put it thus:

> In the future, we shall look upon these regions into which we now penetrate by means of such large telescopes, as a naturalist regards a rich extent of ground or chain of mountains, containing strata variously inclined and directed, as well as consisting of very different materials.

By the end of the nineteenth century, the idea that by studying creation one could learn what was in the mind of the Creator was a common one. Legions of country clerics in Europe and the United States studied the land and life forms around them, becoming experts in their own little areas. Countless sermons and essays were written to prove the existence of God and demonstrate His goodness by studying the shape of the butterfly's wing or the nesting habits of the finch.

It seems to me that Herschel may have been bit ahead of his time in his attempt to fathom the structures of the heavens. Like nearly all pre—twentieth century scientists, he was a devout man. In his study of the skies he was a brother to the clergyman looking at the butterfly wing. By studying the creation, he could learn what was in the mind of the Creator and, if he was lucky, how human beings were to fit into the grand scheme of things. I call this process of studying nature in order to elucidate theological truths "reading the mind of God." It's what Newton, Herschel, and many other early scientists felt they were doing, at least in part, and this goes a long way toward explaining the presence of theological arguments often found in their writings. Today, we do not think of science in this way—we have different goals. But that does not change the fact that many great scientists spent a great deal of time trying to tie their work to theology.

The First Map of the Universe

It wasn't just the lack of interest by other astronomers in studying stars that made Herschel's self-imposed task of finding the structure of the universe difficult. There were a number of technical problems faced by astronomers at that time as well. The most important was the fact that no one had been able to devise a way to measure the distance to any star in the sky. Thus, any attempt to find "construction" was blocked by the inability of astronomers to discover the scale of the universe. Herschel may have wanted to be the natural historian of the heavens, but he was a natural historian who couldn't tell whether a mountain was closer or farther away than a pebble.

The reason for this state of affairs is easy to understand. The stars are very far away. About the only way that someone in Herschel's time could measure the distance to one of them would be to have a telescope capable of such fine resolution that it could see the slight shift in the apparent position of a nearby star that occurs as the earth moves from one end of its orbit to the other. In principle (see fig. 4.1), nearby stars should appear to shift as the line of sight to them passes first near one distant star, then another. By measuring the angles of the lines of sight and using some simple geometry, the distance to the nearby star can be determined.

Despite his lifelong effort, and despite the fact that he built the best telescopes that had ever been available to astronomers, William Herschel was never able to determine the distance to a single star. Indeed, it wasn't until sixteen years after his death that this job was finally done. Thus, Herschel had to approach his task without one of the most important pieces of information needed to accomplish it. In a later section we will talk about his attempts to build even better telescopes than those he had. We shall see how important seemingly grubby skills like metal casting can be to unraveling great questions in science. For the moment, however, let's see how Herschel tried to explore the universe without even a single yardstick to tell him how far away the pieces were.

When you're trying to make progress in the face of ignorance, you have to make assumptions. The assumptions may turn out

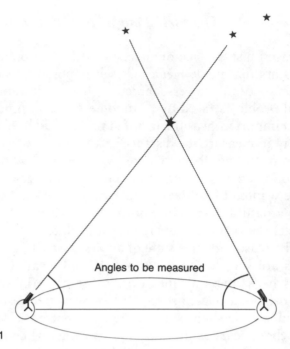

Angles to be measured

FIGURE 4.1

to be wrong — that's a risk you take. But the choice is either make the assumptions or do nothing, and Herschel, like most scientists, chose to take the risk. He started to sweep his telescope across the skies, counting the stars he could see in each direction. He then made three assumptions:

1. All stars have roughly the same intrinsic brightness.
2. There is nothing between us and the stars to absorb light (this assumption was not stated explicitly, but is implicit in his method).
3. The stars are scattered more or less uniformly in space.

All these assumptions are wrong, some more than others. Nevertheless, if we assume the statements are true we can take things like Herschel's star counts and construct from them a map of the universe (or at least of the Milky Way). The idea is this: The number of stars you see when you look in any given direction must depend on the distance between you and the boundary of the stars — the more stars you see, the farther it must be to the edge. It's a tedious job counting all those stars, of course, but the thrust of the logic is simple. Once you know how many stars there are in a given direction, you know how thick the stellar "stratum" is.

In 1785, Herschel presented his first attempt to unravel the construction of the heavens to the Royal Society. A sketch of his Milky Way is shown in figure 4.2. He concluded that the sun was roughly at the center of an amoebalike collection of stars.

He was wrong, of course, as anyone who has seen a photo of a galaxy knows. The main reason lies in his unspoken assumption, 2 above. In point of fact, the plane of the Milky Way is littered with dust and debris that blocks our vision of distant objects. Looking into the galaxy, we are like a motorist driving through a fog—we see ourselves in the center of a fuzzy sphere, and see nothing outside of that sphere. In the same way, Herschel was able to see only nearby stars, and was unable to see what lay beyond them. Nevertheless, the map in figure 4.2 marks the first attempt to come to terms with the larger universe in which we find ourselves. As such, it should be thought of as being analogous to the first maps of the western hemisphere brought back by European explorers. The importance of these maps was not their accuracy, but their existence. They marked the opening of new frontiers.

Uniformity Among the Stars

In the process of scanning the heavens to learn their construction, Herschel made a number of incidental discoveries. From our point of view the most important of these was double stars.

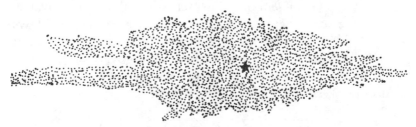

FIGURE 4.2

These are systems in which two stars circle each other in orbit. Today we know that about two-thirds of all stars in the sky are in multiple systems of one sort or another. In Herschel's time, insofar as anyone thought about it, it was generally assumed that other stars were pretty much like the sun — a single point of light in space. Initially, Herschel's observations just showed that there were a lot of stars in the sky that were very close together. Other observers had not been able to discover this fact for the simple reason that their telescopes were not powerful enough — to them Herschel's double stars often appeared as a single, undifferentiated blob of light.

As the years of systematic observation went by, Herschel occasionally would remeasure some regions of the sky. He found that some of the double stars had shifted positions with respect to each other — gotten closer together or farther apart. This is, of course, exactly what you would expect if the two were circling each other under the influence of their mutual gravitational attraction. Thus was Newton's universe extended to the stars.

I haven't been able to find anything written by scientists at the close of the eighteenth century indicating that there was a debate on the question of whether the force of gravity operated among the stars as it does in the solar system. This notion was neither accepted nor rejected; the question simply wasn't asked. My guess is that if someone had asked Newton whether his law would govern the orbits of double stars, he would probably have answered yes, but wouldn't have regarded the question as very important since there was no way to test it against nature.

Thus, the giant step in the development of the principle of universality that Herschel took — extending gravity from the planets to the stars — attracted little attention, either from his contemporaries or from historians. Nevertheless, it forms a vital piece of the development of our present ideas about the universe. As we have seen, there is no logical reason why the same laws should hold everywhere in the universe. That the same force of gravity that keeps the earth in orbit should govern all the stars in the Milky Way is every bit as miraculous as Newton's connection between the apple and the moon.

The Phenomenon of "Magic Hands"

William Herschel was able to open the stellar frontier because of one thing: his ability to build and operate more powerful telescopes than those of any of his predecessors or contemporaries. He was the first example of a phenomenon that has become increasingly common in experimental science—the person with "magic hands."

Look at any great experimental breakthrough in science and the chances are good that somewhere you will find someone with this attribute—someone who has a feeling for the equipment, who can make things work. I once heard the American Nobel laureate I. I. Rabi described as a man who "can just get in there and ride around with those electrons." In this age when the size of experimental groups can get into three digits, when the leaders of those groups spend more time as managers than as researchers, it is perhaps a good idea to recall that somewhere in the scientific enterprise there has to be an individual confronting the hardware alone. If the enterprise is to be successful, that individual had better have magic hands.

That Herschel was one of these people can scarcely be contested. His description of the discovery of Uranus, for example, shows clearly that he knew he had something special the minute the planet swam into his field of view. With the kind of telescope he had at the time, this was no mean achievement. In fact, after the discovery of the planet it was found that other astronomers had not only seen it in the past, but entered it into their journals as a star.

What could have caused such a difference in perceptions? Herschel may have had better eyesight, of course, but there's more to it than that. There are many things that affect the image you see in a telescope: the mirror, the temperature, the state of the atmosphere, to name a few. Obviously, Herschel had a feel—an intuition—about what these things were doing to his instrument on that fateful night. Given this feeling, he was able to recognize that the image he saw wasn't just some sort of glitch, but something that needed attention. This is exactly what magic hands (or perhaps, in this case, "magic eyes") will do for you.

One thing I have noticed about people who have this sort of gift is that they enjoy getting their hands dirty—they like to build the equipment they use. To them the activity is more play than work, in fact.* This attitude has important consequences. There can be little doubt, for example, that the intuition that aided Herschel in his discovery of Uranus was enhanced by the fact that he had built the telescope himself, and knew its idiosyncracies.

Throughout his life, in fact, he insisted on building his own telescopes. He was good at it—indeed, he often supplemented his royal pension by selling his instruments to other astronomers. For them, a Herschel telescope had roughly the same value as a Stradivarius has for musicians.

You can get some feeling for the way Herschel worked by listening to his own description of an attempt to cast a metal mirror for a large telescope in 1781.†

> When everything was in readiness, we put our 537.9 pounds of metal into the melting oven and gradually heated it; before it was sufficiently fluid for casting we perceived that some small quantity began to drop through the bottom of the furnace into the fire. The crack soon increased and the metal came out so fast that it ran out of the ash hole which was not lower than the stone floor of the room. When it came upon the pavement the flags began to crack and some of them to blow up, so that we found it necessary to keep a proper distance and suffer the metal to take its own course.

In fact, other observers tell us that the pieces of flagstone were bouncing off the ceiling, so that Herschel and his helpers (who included a knight and a fellow of the Royal Society) had to run outside to escape.

By repeating this description, I want to illustrate an important point about science, one that is almost never emphasized by commentators. Great questions in science—questions like the ones Herschel raised about the structure of the universe—are

---

* If you'd like to read a delightful account of a modern physicist with magic hands, see *Alvarez: The Adventures of a Physicist* (Basic Books, 1987) by the late Luis Alvarez.
† At this time, telescope mirrors were made of highly polished metal, rather than glass as they are today.

seldom answered by ivory-tower types engaging in pure thought. They are answered by people who are willing to get down into the trenches and grapple with nature. If that means casting your own telescope mirrors, as Herschel did, so be it. Throughout this book we will see examples of important advances in our knowledge of the universe coming about because of advances in technologies that appear, at first glance, to have no relation to the larger question of the universe at all.

In this case, we have already seen that one of the main obstacles to Herschel's survey of the heavens was the fact that none of his telescopes were good enough to detect the very small differences in line of sight to nearby stars associated with the earth's motion in its orbit. The answer to a fundamental question—What is the scale of the universe?—thus depended on the ability of astronomers to build a better telescope. But this task is not glamorous. It depends on one's ability to produce high quality lenses or mirrors, and this, in turn, depends on the details of casting glass (or, in Herschel's case, metal) so that the resulting material produces the minimum amount of distortion in the light that interacts with it. This is detailed, painstaking work. It lacks the grand sweep of the great scientific questions. And yet without it we would have no means of answering these questions. This is one of the great contradictions in the life of a scientist: the most dramatic consequences often flow from the most prosaic sources.

A Summing Up

By time time of Herschel's death in 1824, a new frontier had been opened. The law of nature that had been discovered in Newton's orchard had been extended to the stars. The work of this man extended the arena in which the human mind could grapple with nature from our own solar system to the ends of the universe. As we shall see in the next few chapters, this extension had important and unexpected consequences.

As for Herschel himself, he wrote his own best epitaph nine years before his death. "I have," he said, "looked further into space than any human being ever did before me."

Not a bad ending for someone who started as a second oboe.

FIVE

The Glassmakers
of Benediktbeuren

For now we see through a glass, darkly.
—I Corinthians 13:12

S O MANKIND'S ATTENTION finally turned to a detailed study of the stars. For two millenia they had been little more than unchanging lights in the sky—a backdrop against which the planets and comets moved. Now, for the first time, they were seen as being worthy of study in and of themselves.

It was gratifying that the same law of gravity that makes the apple fall and keeps the moon in orbit seemed to operate in the realm of the stars, but this discovery took scientists only a short way. There are more mysteries in the heavens than those associated with the orbits of multiple star systems. In particular, once Herschel and his successors had mapped out the stellar "strata"—once we knew *where* the stars were—it was only natural to turn to the question of *what* they are. In point of fact, the transition between these two questions occupied much of the astronomy of the nineteenth century. In modern terminology, this period marked the replacement of astronomy (Where are they?) by astrophysics (What are they?).

This shift in emphasis opened a Pandora's box as far as the notion of universality is concerned. Are the stars made of the same stuff as the earth? There is no reason to suppose that they are not, but there is no reason to suppose that they are, either. Why should the chemical elements that make up our little corner of the universe reappear anywhere else? And even if they do, what guarantee is there that they will obey the same laws as their counterparts in our laboratories? The story of how we came to the comfortable conclusion that the earth is typical of the rest of the universe is an interesting one and will occupy us for the next several chapters.

Are the stars made of the same stuff as the earth? To answer this question you have to ask what chemical elements appear in them. But how are you to answer this question, given that you can never get a chunk of star stuff to analyze in a laboratory? The road that eventually led to the resolution of these difficulties began in Munich in 1801, when the French military government was having trouble getting a good survey done of the German territories that Napoleon had only recently acquired.

A Nineteenth Century Silicon Valley

When you see surveyors at work on a highway or a building site, you notice that they use a three-legged instrument called a transit. The transit is really a small telescope used to sight over long distances, affording straight lines of sight over the uneven earth.

At the heart of any transit is a glass lens. And it was here that the military government's problem lay. A good lens requires high-quality glass, and in 1801 there was precious little of this commodity being produced. Even worse than this shortage, at least from the point of view of Napoleon's representatives, was the fact that the main center for high-quality glass production was in London.

As so often happens in high-technology enterprises, the flourishing of the London manufacturing centers was the direct result of the work of one man. In this case, the man was John Dolland, who, in 1758, developed a lens system which would bring all colors of light to a focus at the same point. This was an extremely important discovery — in fact, there is a descendant of his so-called achromatic lens in every camera now in existence. The result of his discovery: what we would call a small high-tech industry centering around the manufacturing of high-quality optical instruments (including astronomical telescopes) and, along with this manufacturing, a small but important industry devoted to the production of high-quality glass.

If you have ever been in an eighteenth-century building and looked out of the window, you know why glass quality is so important in optical instruments. The images you see from inside the building are wavy and distorted. The straight edge of a neighboring building may be turned into a fun-house curve. The glass seems to be of uneven thickness, and numerous bubbles make things even worse. It's hard to see the true shape of large objects on a sunny day through such glass, so you can imagine how hard it would be to look at a pole held by a distant surveyor. Small wonder, then, that as the demands placed on optical instruments became more stringent, a source of high-quality glass for those instruments had to be found. And small wonder that with so many military and civilian uses for the new optical technologies, nineteenth-century versions of Silicon Valley began to spring up around the European continent.

Glass has been made since ancient times by the process of mixing sand and other materials, melting them, and then pouring the resulting liquid into molds. The primary difficulty in the manufacturing of old glasses was the failure of the artisans to get all of these ingredients to mix together into a uniform melt. The resulting glass would actually have a different composition in one part than in another, like a poorly mixed salad. This is one way that light coming through an old window pane is distorted. Bubbles of trapped air are another. Making a glass that is free from these defects is not a glamorous occupation for a scientist now, nor was it one in the early nineteenth century. It is a perfect example of one of those grubby tasks on which the progress of science depends. And as often happens, the job was not done by a research scientist, but by an artisan and industrialist. In the process, a German craftsman by the name of Joseph Fraunhofer achieved the species of immortality that comes to someone who has an important discovery named after him.

The Fraunhofer story actually begins with the realization by a group of venture capitalists in Munich that the problems their local government was having with surveying instruments represented a major commercial opportunity. One of them, a man by the name of Joseph Utzschneider, was already a major figure in the cloth and leather industries. He instituted an international search to find someone at the forefront of glassmaking and to bring him to Germany. He found Pierre Guinand, a Swiss who had devoted over a decade to experimenting with ways of producing a uniform mixture in molten glass, and brought him to Bavaria to build a factory. The firm, called Mathematical Mechanical Institute Reichenbach, Utzschneider, and Leibherr eventually set up operations in an old Benedictine monastery about thirty miles from Munich.

The monastery, called Benediktbeuren, had a very interesting history. It was founded in the eighth century by St. Boniface, then burned by the Hungarians in 985. It was rebuilt by the Benedictines in 1031 but sacked by the Swedes in 1632, during the Thirty Years War. By 1805 it had fallen into disuse and was available for conversion into a glassworks. It was ideal for the purpose, since it possessed several large buildings and was located in a forest, so that abundant firewood was available.

The new technology—what would make Benediktbeuren the Silicon Valley of its time—was a technique for mixing the molten glass while it was still in the crucible, thereby insuring a uniform quality on cooling. The problem that had been solved is simply stated: How can you mix the molten material without having the mixing paddles contaminate the glass or introduce bubbles? The solution that Guinand hit upon involved mixers made of firebrick—the kind of brick used to line fireplaces and furnaces. For fifteen years he experimented with this technique, until he was able to produce large melts whose uniformity of composition was better than anything that had ever been known.

Simple as it may sound, this was the foundation on which the German optical industry built its position of world leadership, only recently challenged by the Japanese.

So important were the secrets of the glassmaking techniques to the company that Guinand's contract stipulated that only he and his wife were to work the furnaces. In addition, he received a large yearly sum to bind him to secrecy but retained the right to instruct one of his sons in the art. In this way there would be someone who could take over if Guinand died. The quest for secrecy even extended beyond the grave, for the contract assured a yearly pension for his wife provided she kept quiet about the procedure.

All this may sound excessive today, but there were no patent laws in the early nineteenth century. And as we know, high-tech companies today are frequently targets of industrial espionage as competitors try to steal the secrets that give a leading firm its competitive edge. I suspect modern Silicon Valley contracts will look as strange to future generations as the Benediktbeuren contracts look to us.

It was during the early period of the new firm that Joseph Fraunhofer came on board as assistant to the man in charge of making optical instruments from the new glass. The eleventh child of a master glazier, Fraunhofer was orphaned at the age of eleven and apprenticed to a glass cutter. At the age of fourteen the building in which he was working collapsed, trapping him under the debris for several days before he was rescued. This drama brought him to the attention of the public and, more important, of the Elector of Bavaria, who awarded the boy the sum of eighteen ducats. This was enough to buy him a book on optics, a glassworking machine, and a release from the remainder of his apprenticeship. After a few years as a journeyman glassworker and a few failed business enterprises, he was ready for the kind of security that the Mathematical Institute offered.

He must have been a fast learner, because by 1809 there was a contract calling for Guinand to teach him the secrets of glass-making. By 1811 he was a full partner in the firm, and in 1813 Guinand packed his bags and returned home, leaving the whole operation in the hands of the young Fraunhofer. Thus, at the age of twenty-six, Fraunhofer was in a position of authority in a major new manufacturing enterprise. This must have surprised his contemporaries, but we have seen so many whiz kids becoming millionaires before their thirtieth birthday that it scarcely seems worth remarking.

Once he had the factory at his disposal, Fraunhofer quickly began to improve his product. He devised a furnace that could bring the melt to high enough temperatures so that bubbles could be eliminated from the glass. He designed and built a series of machines for making lenses, prisms, and other optical components of hitherto unknown quality. Like Herschel, he was a man with magic hands—a man who paid a great deal of attention to the details of his work. We know something about the way he worked from his journals. Here is an extract from his account of the mixing of a batch of glass in 1814.

> On Jan. 20 we brought the pot to the furnace. The bottom of the outer pot was covered with clean crushed quartz to a depth of about 1″ and the inner pot placed inside. The pots were then hung by three irons to the rope of a large crane. The bottom of the furnace was covered with 1½″ of clay and the pot lowered onto it. On the 21st a fire of undried wood was started. At night the inspection hole was closed and in the morning a fire made again; until the morning of the 26th this procedure continued. On that night dried wood was charged, so that by the morning of the 27th the fire was hot enough. . . . At 9:30 filling of the batch commenced. Only two shovelfuls were filled on at a time, so as not to chill the pot excessively. At 7 in the evening the filling was finished. From 11 midday to 9 in the evening the glass was stirred every 8–10 minutes. At one o'clock (on the 28th) the cover of the furnace was removed and the glass carefully skimmed, and at 1:45 the fires again were stoked. At 8:30 in the evening preparations were made for the last stages, and at 10 stirring began again. The glass was fairly fluid. At 1:30 A.M. on the 29th stirring ceased. The furnace remained open for an hour and was then closed to slow the rate of cooling.

Like Herschel before him, Fraunhofer was not a man to leave the heavy work to others. When you realize that cooking up a single batch of glass took almost two weeks of work, day and night, you begin to get some idea of why the development of high-quality glass was such a long time coming. I hate even to think about all the things that could have gone wrong with the melt Fraunhofer describes. But this attention to detail, this willingness to be there at 1:30 in the morning to supervise a crucial moment, paid off. Within a few years, Fraunhofer was producing the best "philosophical and mathematical instruments" anywhere in the world.

Gauging Sunlight

But as he pushed into new regimes of quality and precision, he ran into a serious set of problems of a type encountered by every technological pioneer. He had instruments that could produce sharper images and see more detail than any that had ever been built. But he had no idea of what those fine details were—of what his instruments were actually doing to the light. One of Fraunhofer's first tasks, once he got the glassmaking under control, was to make some precise measurements on the way his new instruments affected the light coming through them. We can take a simple glass prism as an example of the type of optical component whose properties Fraunhofer had to determine.

Isaac Newton had demonstrated that if sunlight is passed through a glass prism, it is broken up into all the colors of the rainbow, from red to violet. If the colored beams are then passed through another prism, as shown in figure 5.1, they form a beam of white light. From this experiment, Newton concluded that white light is a mixture of all colors.

The reason that Newton was able to do this experiment is that the glass in the prism bends different colors of light by

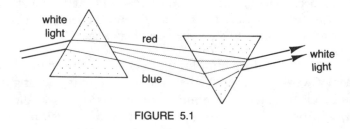

white
light

red

blue

white
light

FIGURE 5.1

different amounts, with red light being bent less than the blue. It is the differential bending of the colors that gives the prism its power to "disassemble" a beam of white light. The amount of bending and the spread of the colors depends on the composition of the glass. Since prisms are often used to bend light beams around corners in optical instruments of various kinds, it was crucial for Fraunhofer to be able to make precise measurements of the amount of bending associated with each fine gradation of color.

To appreciate the sort of problem Fraunhofer had to solve, consider what must have been a typical situation. Suppose he was selling two instruments—one to a laboratory in Sweden, the other to a laboratory in Spain. It would have been extremely inconvenient for people from either laboratory to come to Bavaria to calibrate their instruments. Suppose further that the two instruments contained prisms that were made from different batches of glass. Because of minute differences in the composition of the glasses, the two prisms would bend incident light through slightly different angles. How could Fraunhofer instruct his clients so that these small but unavoidable differences wouldn't make it impossible for them to compare measurements?

The standard way to solve this problem is to have people in the two laboratories measure the same thing and compare readings on their instruments. For example, if they were using thermometers instead of prisms they could both measure boiling water and see how much the readings of their thermometers differed. This would tell them how to correct for differences in the instruments.

What Fraunhofer needed, then, was the optical equivalent of boiling water—something that would be the same at both laboratories, so that any differences in readings could be ascribed to the instruments themselves. Light from the sun is an obvious choice for such a use—the same sun shines in Spain and Sweden. He couldn't just choose a color in the solar spectrum, however, because it was very difficult to specify exactly which color to pick out of the smooth gradations from red to violet. What he needed was a set of markers in sunlight.

Fortunately, the superb quality of his instruments allowed him to use a discovery that had been made in 1802 to "mark" sunlight for him. It had been noted by the British chemist Henry Hyde Wollaston that superimposed on the spectrum of light from the sun were a series of narrow, dark bands of the type sketched in figure 5.2. Fraunhofer had no idea what these lines were or why they were there, but for his purposes such knowledge was unneccesary. All he needed to know was that an irregularly spaced, easily reconizable pattern of these lines was present in sunlight. These bands could serve to "mark" the spectrum for him.

Suppose, for example, that he found that the colors around the complex of lines labeled 1 in the drawing were bent through one angle by one prism, but through a slightly different angle by another. He would then know how to calibrate the instruments, and, what is more important, he could tell his clients in Spain and Sweden how to do the calibration themselves. It would just be a matter of pointing the instrument at the sun and seeing the angle through which that particular complex of lines is bent. Comparing this to the angle of bending associated with the angle measured in the other would allow both users to be certain that their instruments were "in tune."

By the 1820s, Fraunhofer had observed and measured no fewer than 574 of the dark lines in the spectrum of the sun. Because he had such fine glass, he was able to see these lines clearly enough to determine their properties. Newton's prism distorted the sunlight so much that the dark lines were simply washed out, while Wollaston's observations were only marginally better. To honor Fraunhofer's prodigious effort in precision pioneering, the lines were named in his honor, and we still refer to them as Fraunhofer lines.

It detracts in no way from his achievement to note that he was not the first to see the lines that now bear his name, nor

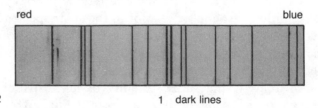

red   blue

FIGURE 5.2   1   dark lines

to remark that he was not the one who finally understood what the lines are. He didn't need to understand them. For him, they were just useful markers that allowed him to carry out his work as a designer of optical equipment.

By using the lines as benchmarks for the optical instruments used by scientists around the world, he brought them to the attention of the scientific community and laid the groundwork for a major advance in our knowledge of the stars. For as we shall see in a moment, the dark lines in the spectrum of the sun (and other stars) are nothing less than the key to the problem we posed at the start of the chapter. They provide us with a way of finding out the chemical composition of objects that, like the stars, we can never touch.

Before we move on to the composition of the stars, however, let's take a moment to finish our story of Joseph Fraunhofer's life. In addition to his traditional optical instruments, he pioneered the use of a device called a diffraction grating, a device that became one of the workhorses of modern astronomy. He also made telescopes that detected the apparent annual motion of stars—the goal that had eluded Herschel throughout his life. He was awarded an honorary professorship in Munich and seemed to be well-launched on a career of scientific eminence. At the age of thirty-eight, however, he contracted tuberculosis and died a year later. On his deathbed, he passed on the secrets of his glassmaking to a colleague, thereby preserving the tradition of his profession.

Spectroscopy—The Path to the Stars

For a full half century the dark lines in the sunlight remained little more than useful benchmarks for practising opticians. Their secrets remained buried while scientists accumulated the skill and knowledge needed to exploit them. Then, in 1859, two German scientists (each well-known for other reasons) got together in a laboratory in Heidelberg and discovered the connection between the lines and the chemical composition of the stars.

Gustav Kirchhoff (1824–1877) was a physicist best known today for his enunciation of the rules that allow scientists to calculate the voltages and currents in even the most complex electrical circuits, Kirchhoff's Laws. His associate, Robert Bunsen (1811–1899), was one of the leading chemists of his time, well-versed in the newly developing arts of purification and identification of chemical elements and compounds. For the average reader, he is probably best known as the inventor of the Bunsen burner, that ubiquitous presence in freshman chemistry laboratories. A somewhat less well-known, but no less significant, chemical achievement of Bunsen's was the discovery of the antidote to arsenic poisoning.

In the mid-nineteenth century, fifty years after John Dalton had first enunciated our modern theory of the atom, chemists were busily engaged in the job of analyzing the composition of all sorts of materials. The tools they had at their disposal were those of what we now call analytic chemistry. Identification involved separating different chemicals by heating, mixing, precipitating, and all the other techniques most of us associate with chemistry. These techniques have some important drawbacks. For one thing, they require that you have the material in hand before it can be analyzed. This makes them useless for dealing with the question of the composition of stars. Furthermore, they require that you have a fairly sizable amount of the material available to analyze—you don't want it getting lost in your flask.

Finally, and perhaps most important, the classical techniques of chemistry require the sample to be analyzed be of extremely high purity. You have to be sure that when you detect the presence of an element that it is actually in the material you're measuring, and not some sort of contamination. The ability to produce samples of high purity was a skill that Bunsen had developed throughout the early years of his career. In fact, he developed his famous burner, with its almost invisible flame, as a tool that would allow him to identify metals and salts by the color of their flames.

These are everyday examples of the sort of thing that was motivating Bunsen's work. Ocean driftwood, for example, is highly prized as a fuel for campfires because it produces multicolored flames when it burns. The colors come from salts absorbed by the wood during its stay in the water—each type of salt imparting a different color.

Similarly, modern street lamps come in two varieties. Those that give a yellow light (commonly used at freeway interchanges) contain sodium vapor. The color is associated with the element sodium, and the lamps are equipped with what are usually called "sodium vapor bulbs." On city streets, on the other hand, the lamps give off a bluish light. This color is characteristic of the element mercury, which, in vapor form, is in the bulbs. Thus, different chemical elements can be associated with the different colors of light they emit when heated.

Actually, by 1859 Bunsen had already taken this process one step farther by passing light from heated (and very pure) samples of materials through an instrument containing a prism. When he did so, he found something unexpected. Instead of a smooth continuous spectrum running from red to violet, he found a series of sharp lines, each associated with a different color, as sketched in figure 5.3. We will discuss the significance and origin of this so-called discrete spectrum in some detail in the next chapter, but for the moment we simply note that while all elements produce such spectra, the details of brightness and location of the lines vary from one element to the next. It had also been known for some time that the bright yellow lines emitted by the element sodium—the lines that give those freeway lamps their characteristic color—corresponded to dark Fraunhofer lines in the spectrum of the sun. This was the state of knowledge in 1859.

Kirchhoff was using this coincidence of bright and dark lines for the element sodium to calibrate an instrument by passing first sunlight and then light from a sodium flame through a

red   blue

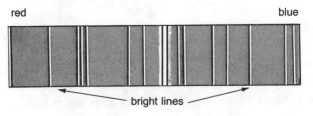

← bright lines →   FIGURE 5.3

prism and adjusting things so that the corresponding bright and dark lines overlapped. He noticed that if he let sunlight into his apparatus, but interposed the sodium flame in the beam, the Fraunhofer lines became much darker and more noticeable than they were in the sunlight alone.

Mulling this unexpected finding over through the course of what must have been a sleepless night, Kirchhoff realized that it could be understood only if the light emitted by atoms of sodium was at exactly the same position in the spectrum as the light which sodium vapor absorbs. What an atom emits, in other words, it will also absorb. Kirchhoff's argument was simple: At the center of the sun was an incandescent core, emitting light of all colors. On its way out of the core, this light passes through the outer atmosphere of the sun, where the temperatures are cooler and some ordinary sodium atoms are to be found. These atoms, by a process we will discuss later, absorb certain colors from the light. The resulting lower intensity of these colors is seen as a dark line in the solar spectrum—the lines used by Fraunhofer to calibrate his instruments. If the beam goes through another region that contains sodium, as in Kirchhoff's experiment, then even more absorption takes place and the dark lines get darker.

This meant that the dark Fraunhofer lines can be interpreted as evidence for the existence of sodium atoms in the sun. They are, in effect, atomic fingerprints whose presence signifies the presence of the atom itself.

It is hard to overstate the importance of this almost accidental discovery. It provided scientists with a way of determining the chemical composition of objects which were either too far away or available in too small a quantity for this determination to be done any other way. During the weeks that followed, Kirchhoff and Bunsen worked feverishly to develop techniques for determining chemical compositions by looking at emitted and absorbed light.

A typical instrument for this sort of work is shown in figure 5.4. Light from some source (a distant star or a burning sample) is brought into the apparatus through a series of slits and lenses, then run through a prism or similar device that will spread out the individual colors. The spread-out beam is then sent out

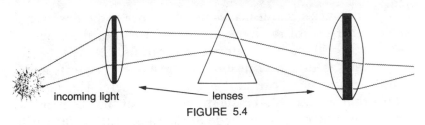

incoming light ← lenses →

FIGURE 5.4

through another set of slits and lenses to the point where it can be observed. In Kirchhoff and Bunsen's time, this observation was done by eye. Later, photographic film was used to record the position of the bright and dark lines. Today, the observation process is done electronically, with analogues of TV cameras. In any case, this instrument is called a *spectroscope* and the branch of science devoted to studying its use is called *spectroscopy*. And while today a spectroscope may come with an on-board computer and cost hundreds of thousands or even millions of dollars, Kirchhoff and Bunsen built their first model out of a couple of discarded telescopes and a cigar box.

The two scientists were not slow in exploiting their new method of analysis. Their ability to analyze very small samples led over the next two years to the discovery of the elements cesium (from the latin *caesius*, or sky blue) and rubidium (from *rubidius*, or dark red) in mineral water taken from a German spa. In both cases the elements were named for the color of the most prominent lines in their spectra.

In our time, the use of spectroscopy has become so commonplace that it is scarcely noticed. The ability to determine the chemical composition of a material by the way it absorbs or emits light is extremely useful in industrial situations, where it is used to monitor and control the manufacturing of everything from structural steel to pharmaceuticals. And when you read about a criminal trial in which microscopic bits of fabric or paint are traced to a given location, it was probably spectroscopy that gave the investigators their information.

But for our investigation of the principle of universality, there is another aspect of spectroscopy that is important. The key point is that once a substance has emitted or absorbed certain colors of light, it is immaterial how far that light has to travel before it is measured. It makes no difference if the observer is at the other end of the room or the other end of the universe —the light still retains its atomic fingerprints.

This means that we can look at light from distant sources and determine what chemical elements are there. This in turn, means that we can ask the questions posed at the beginning of this chapter: Are the stars made of the same stuff as the earth, and do atoms behave the same elsewhere as they do here? It is to these questions that we will turn in the next few chapters.

Before leaving the pioneers of spectroscopy, however, I would like to make one more point. In the early nineteenth century the French philosopher Auguste Comte, the founder of the new science of "social physics" (now called sociology), published a list of what he called unanswerable questions. Prominent on that list was the question of the chemical composition of stars —a question we now propose to answer using the discoveries of Fraunhofer, Kirchhoff, and Bunsen. Comte's contention, of course, was based on the assumption that you have to have a piece of material in your hands before you can analyze it. He couldn't have foreseen the development of spectroscopy. But isn't it marvelous that the work of a man toiling away over a furnace in a remote Bavarian monastery should open a path to the stars—a path that prominent philosophers said couldn't exist?

SIX

Atoms and Light

For in and out, above, about, below
'Tis nothing but a magic shadow show.

—OMAR KHAYYÁM
The Rubáiyát

THE DEVELOPMENT OF spectroscopy opened new vistas for scientists who wished to study the universe in the large. It was no longer necessary for them to restrict their attention to objects that could be brought into their laboratories —they could determine the chemical composition of stars at the far end of the universe. As Kirchhoff realized, the dark Fraunhofer lines in light striking the earth are shadows cast by atoms—shadows that can be seen from one end of the universe to the other.

If we are to understand how to interpret these shadows, we have to understand something about the nature of light, the nature of atoms, and the nature of the interaction between the two. I won't trouble you with a long account of the historical struggle scientists went through to gain this understanding. Instead, I will give you a short explanation of our current thinking

71

on these subjects. If you already know about them, or if you are impatient to continue with the discussion of the idea of universality, you can skip the details and move ahead to the summary at the end of the chapter.

Light

Light is a wave. As such, it is a cousin to things like surf coming into a beach or a vibrating guitar string. All waves share one characteristic: The motion of a wave is not the same as the motion of the medium in which the wave moves. When surf comes in to a beach, for example, the waves come in straight, but the water on which the wave moves goes up and down. If you've ever swum out in the surf, you have direct evidence for this statement—as the wave goes by, it lifts you up, then drops you down again. It does not carry you with it to the beach.

This distinction between the wave and the medium on which it moves is important, for it tells us that the wave is different and distinct from that medium. A light wave is the ultimate realization of that distinction, for it is a wave that requires no medium at all. This is why it can travel through the vacuum between the stars and the earth.

In figure 6.1 we sketch the modern idea of the internal structrure of a light wave. It consists of electric and magnetic fields which are perpendicular to each other and to the direction in which the wave moves. These fields are a description of the forces that the light wave exerts on objects it passes, just as the surf exerts a force on you when you're in the ocean. When a light wave moves by, an object with an electrical charge (such as an electron in an atom) will move up and down.

An important way of characterizing any wave is a quantity called the *wavelength*, which is defined as the distance between successive crests of the wave. For surf at a beach the wavelength might be 10 to 20 feet. For light, the wavelength is much shorter—roughly 0.001 millimeters (the size of a fine particle of smoke in the atmosphere). Another way of characterizing the wavelength of light is to say that in the space between crests you could fit several thousand atoms side by side.

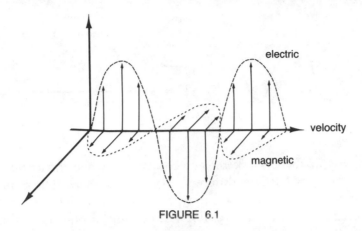

FIGURE 6.1

The most striking property of light—its various colors—is related to the wavelength. The general rule is that the bluer the light, the shorter the wavelength. Between the crests of a light wave that we would percieve as being red you can fit almost eight thousand atoms. As you moved through yellow and green toward blue and violet, this number would drop until, for the bluest light the eye can see, it would become about four thousand.

The Photon

In the nineteenth century, physicists thought of light in much the same way as they thought of waves on the ocean. The chain of crests and troughs was held to be, for all practical purposes, infinite in extent. The waves, in other words, were continuous. In this century, we have developed a new way of thinking about light and atoms—a way of thinking embodied in a science called *quantum mechanics*. One of the principle tenets of this science is that at the atomic level nothing is continuous. Everything, even light, comes in bundles of matter and energy called *quanta* (singular: *quantum*, the Latin for "so much").

The quantum of light—the bundle in which it comes—is called a *photon*. It may help you to remember the term if you think of the "photon torpedoes" in the old *Star Trek* adventures.

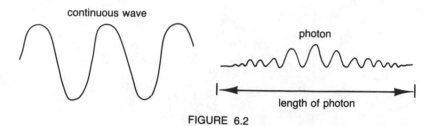

FIGURE 6.2

In figure 6.2 there is a sketch of a continuous wave and a photon, so that you can see the difference between the two. The photon is analagous to a tidal wave in water, for it has a definite length, denoted by L in the figure. For visible light, a photon is about 3 feet long.

Incidentally, this picture of the photon goes a long way toward explaining one of the most difficult conceptual problems that newcomers to quantum mechanics experience—the so-called problem of wave-particle duality. There are experiments you can perform with light in which the photons seem to act just like surf at a beach (i.e., like a wave) and other experiments in which they seem to act like baseballs (i.e., like particles). From the sketch, you can understand a little of how such a situation could arise, for the photon is localized in space (like a particle) but has a reasonably well-defined distance between successive crests (like a wave).*

The bundle of energy we call the photon moves through space, losing no energy as it goes. The rules of quantum mechanics tell us that the energy it carries is related simply to the wavelength of the light—the shorter the wavelength, the higher the energy. Since red light has a wavelength almost twice as long as blue, a photon of red light will have half as much energy as a photon of blue light.†

There is a rough kind of confirmation of this notion in everyday experience. If you put a piece of metal into a fire, it will start to glow as it heats up. The first color you will see is red,

---

* A more complete discussion of this problem is given in my book *The Unexpected Vista* (Scribners, 1983).

† I remind experts that the precise relation is $E = hc/\lambda$ where lambda is the wavelength of the light.

corresponding to the lower energy photons of visible light. When the metal is white hot, it is emitting light of all wavelengths. It has to be at a very high temperature before it emits photons at wavelengths corresponding to blue light and becomes blue hot.

The Atom

Now that we understand something about the nature of light, we can turn to the question of how it is created. Since light originates in things like lamps and stars, and since those objects are made of atoms, it follows that if we are to understand the origin of light, we are going to have to understand the nature of the atom.

The atom consists of a small, dense, positively charged nucleus around which much lighter, negatively charged particles called electrons move in orbit, much as the planets orbit around the sun. Unlike the planets, however, electrons cannot exist in orbits at any arbitrary distance from the nucleus. There are only a few so-called allowed orbits for the electrons. Each of these orbits is at a different distance from the nucleus, as is shown on the left in figure 6.3.

You can think of the electron in the atom as being in a situation analogous to a man on the steps of a stadium. He can remain standing on one step or he can move up or down one or more steps. But under no circumstances can he stand between steps. In the same way, the electron can be in one allowed orbit or another, but it cannot be between them.

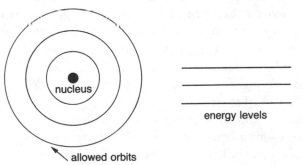

nucleus

energy levels

allowed orbits

FIGURE 6.3

READING THE MIND OF GOD

The man in the stadium has to expend energy to move up the stairs, and he gains energy as he moves down. In the same way, energy has to be added to an atom if the electron is to "climb the stairs" and move to an upper orbit. An electron left to itself in an upper orbit will "roll down the stairs" to a lower orbit. This means that we can think of each allowed orbit as a state corresponding to a specific energy of the electron. On the right-hand side of figure 6.3 we convert this realization into a drawing, substituting a series of ascending energy levels for the series of concentric orbits shown on the left.

The picture of the atom just described was first developed in 1912 by Niels Bohr, a young Danish theoretical physicist working in Manchester, England. For this work he was later awarded the Nobel Prize. Today, one often hears this picture of the atom with allowed orbits referred to as the "Bohr atom."

Absorption and Emission of Light

Quantum mechanics tells us that when an electron in an atom changes orbit, it does so in a *quantum leap*. It disappears from one orbit and reappears in the other without traversing the space in between. (Don't try to picture this—it can't be done.) If the electron is "rolling down the stairs" to a lower orbit, the energy difference between the initial and final states is emitted by the atom in the form of a photon. This process is shown diagramatically on the left in figure 6.4. If, on the other hand, the electron is to "climb the stairs" from a lower to an upper level, then the energy needed to drive the transition must be absorbed by the atom. This energy can come in the form of an absorbed photon (as shown on the right of fig. 6.4), through a collision with another atom, or through other, more complex processes.

Once we have this picture of the relation between photons and the energy levels of atoms firmly in mind, we are in a position to understand how we can do chemical analyses of the stars. The best way to start is to point out that pure samples of chemical elements emit only certain colors of light.

For the sake of simplicity, suppose that electrons in a given atom have only three allowed orbits, as shown in figure 6.5.

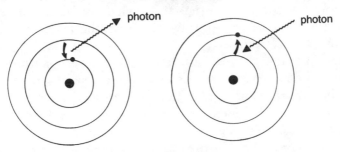

FIGURE 6.4

Suppose further that when a collection of atoms are heated the resulting collisions add energy to a given atom so that an electron finds itself in the uppermost orbit. That electron will spontaneously "roll down the stairs" to the bottom level, of course, but there are two ways it can do so. It can make the jump all at once, or it can make two jumps — the first to an intermediate level, the second from there to the bottom level.

The point is that each of these three jumps corresponds to a different energy for the photon that is emitted when the jump occurs. From the discussion above, we can see what this means: Each jump involves a different energy difference and hence a different color of emitted light. When there are many atoms in a sample, some atoms will take the first path outlined above, others the second. The result will be that all three colors of light will come out of the material in which the atoms are found. If this light is passed through a prism, we will see the pattern shown on the right in the figure. This pattern is precisely the sort of thing Bunsen saw in his work — what we called a discrete

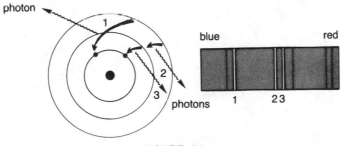

FIGURE 6.5

77

spectrum. Of course, in a real atom where hundreds of energy levels are available to electrons, there will be more than three lines in the spectrum.

The fact that each chemical element is made up of atoms with a different number of electrons and different sized nuclei means that the arrangement of the energy levels in the atoms of one element will never be exactly the same as those for atoms in another. Thus, no two elements will emit photons with exactly the same energy. This in turn accounts for the fact that when emitted light is passed through a prism, no two species of atoms will produce exactly the same lines in the spectrum. This is how it is possible to identify different chemical elements by the light they give off. The spread of bright lines characteristic of a given type of atom is called an *emission spectrum*.

Turn next to the inverse process—one in which atoms absorb, rather than emit, light and consider for the sake of simplicity the atom with only three levels. Suppose that a beam of white light is directed on a collection of such atoms. What will the light that emerges from the other side look like?

White light, by definition, contains photons corresponding to all wavelengths. If most of the electrons in our imaginary atoms are in the innermost orbit, two things can happen, as shown in figure 6.6: The electron can absorb a single photon, rising in a single step to the uppermost level, or it can absorb two photons—one to get it to the intermediate level, another to get it to the top. The result will be that three types of photons will be absorbed from the incoming beam—one corresponding to each of the three energy level differences in the atom.

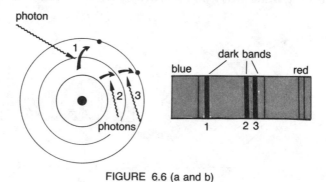

FIGURE 6.6 (a and b)

Once the electrons are lifted up by the absorption process, they will eventually fall back down. But when photons are emitted from the atoms during the fall, they can be emitted in any direction. There is a very small chance that the photon emitted by an atom will be moving in the same direction as the incoming beam. This means that someone looking at the white light as it comes through the atoms will not see the reemitted photons. What will be seen is white light with certain colors removed. The spectrum seen by the observer is shown on the right of the figure. It should be obvious that these dark bands are precisely what we labeled Fraunhofer lines in the last chapter. It is also what Kirchhoff saw on that afternoon in 1859.

The final result of the process by which atoms remove certain wavelengths of light from a beam is called an *absorption spectrum*. By the same reasoning as that above in the case of the emission spectrum, each chemical element will have a different pattern of dark lines in its absorption spectrum. An atom can be identified as easily by what it absorbs as by what it emits.

The correspondence between the bright lines in an emission spectrum and the dark absorption lines that so puzzled Kirchhoff and Bunsen can now be understood very easily. The lines correspond exactly because they are produced by transitions between the same energy levels in the atom. They differ only in the direction of the transition. Thus, the wavelength associated with emission (a downward transition) must be exactly the same as that associated with absorption (an upward transition).

Looking at light coming from distant sources can thus provide us with two distinct types of information. It can tell us what sorts of materials are being heated to give off the light, and it can tell us something about the medium that the light has passed through on its way to our instruments. Modern astronomers routinely use both techniques in their work.

Further Important Details About Spectroscopy

MOLECULES

Up to now we have talked only of photons emitted by atoms in quantum jumps of single electrons. There are other processes that can produce photons, and the most important from our point of view are those that occur in molecules made up of two or more atoms. In such systems electrons can make the usual kinds of jumps within individual atoms, but other kinds of motion can occur as well. For example, the atoms in a molecule can rotate around each other, or they can vibrate as if they were held together by springs.

When we apply the laws of quantum mechanics to a description of molecules, we find that rotation and vibration produce energy levels similar to those associated with the orbits of individual electrons, even though they involve a complicated cooperative motion of all the particles in the molecule. As was the case for individual electrons, the molecule can move from one energy level to another—for example, from a slow vibration to a fast one—only by emitting or absorbing photons with the appropriate energy.

This means that we can detect the presence of molecules like carbon dioxide and ammonia or their more complex cousins in distant sources of light, just as we can detect the presence of single chemical elements.

IONS

As the temperature of a material goes up, the increasingly violent collisions between atoms start to tear electrons loose from their moorings. We say that the atoms become ionized. The loss of one or more electrons from an atom in the ionization process affects the emission and absorption spectrum.

The reason for this is that while the most important force exerted on a given electron in an atom is the attraction to the nucleus, that electron, because it carries a negative charge, is also repelled by the other negatively charged electrons in the atom. The energy level of the electron depends on both these forces. This means that if an electron is removed from an atom, the energy levels of the remaining electrons will shift around a little in response. As a result, the energy (and wavelengths) of the light emitted by an ion will be different from that emitted by an atom with its full complement of electrons.

This is an interesting point, because the amount of ionization that a material suffers—the number of electrons lost per atom—depends on the temperature of the material in which the atoms and ions reside. By looking at the spectra from a distant star, then, we can estimate the temperature at the surface of that star.

FINE STRUCTURE

As spectroscopes became more and more precise during the latter part of the nineteenth century, an unexpected development occurred. When examined under high levels of resolution, single lines in emission and absorption spectra were seen to be made up of sets of multiple, closely spaced lines. This phenomenon was known as the appearance of fine structure in the spectrum. Then, under even more magnification, some of these multiple lines were seen to be multiple themselves. This progression is shown in figure 6.7 below.

Once again, we resort to the laws of quantum mechanics for an explanation of these effects. They tell us that what we have up to now considered a single allowed orbit in an atom will, upon close examination, be found to correspond to a set of closely spaced energy levels. For example, if we have two electrons in a given orbit, one traveling clockwise and the other counterclockwise, quantum mechanics tells us that the energies of the two electrons will be slightly different. Thus, a single energy level (corresponding to the orbit) would "split" into two   one for each direction of motion. As shown in figure 6.8, page 82, if two energy levels undergo this sort of splitting, a single line in the spectrum of the atom (corresponding to the single emitted photon on the right) would be transformed into

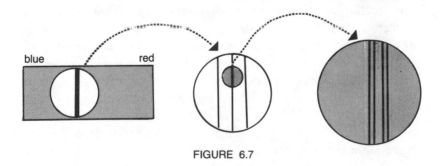

FIGURE 6.7

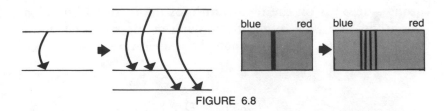

FIGURE 6.8

four separate lines (corresponding to the four possible quantum jumps on the right).

The energy levels in any atom are always split, of course, but it you look at the spectrum with an insufficiently precise instrument, all the multiple lines merge together and you see only one line. This would lead you to conclude (falsely) that there was no splitting of the levels. You would see the true state of things only if you replaced the poor instrument with a better one—one that could separate the four lines. This is exactly what happened to scientists as their instruments improved throughout the latter half of the nineteenth century.

The explanation of the second, finer splitting of the spectral lines is similar to, but more complex than, the explanation of the appearance of fine structure. It turns out that both the electrons in their orbits and the nucleus of the atom share a simple property: Both can be thought of as spinning around an axis, much as the earth spins around a line drawn through the North and South poles. Thus, an electron moving in a counterclockwise orbit at a particular distance from the nucleus can be spinning around its own axis in either a clockwise or counterclockwise direction. Similarly, the nucleus itself can have its spin oriented in a number of possible "allowed" directions. The rules of quantum mechanics tell us that each possible combination of the respective spin orientations represents a state with a slightly different energy than any other combination. Thus, the spins of the particles lead to a further splitting of energy levels along the lines outlined above for fine structure. And as before, this splitting of energy levels is mirrored in a splitting of lines in the emission and absorption spectra of the atom.

The details of how this final splitting of the lines comes about depends on the way the atom is constructed: It varies from one atom to the next. Still, it is fair to say that scientists today understand most of the details of the spectra we see in our laboratories and telescopes.

The fact that electrons can be thought of as spinning around an axis leads to another important property of spectra. If the electron happens to find itself in a magnetic field (as it would, for example, if it were circling an atom on the surface of the sun), its energy will depend on the strength of that field and on whether the spin is clockwise or counterclockwise. The stronger the field, the greater the difference in energy between electrons that have different directions of spin.

On the left in figure 6.9 we show the splitting of an energy level by a magnetic field. The initial energy level is split into two closely spaced lines, but the splitting is much greater for the stronger field.

This property of the electron leads to an important conclusion: If we measure the splitting of the spectral line, we can deduce the splitting of the energy levels and hence the magnitude of the magnetic field that is causing the splitting. So by carefully measuring the spectrum of an atom, we can measure the magnetic field at the atom's location, even if it is billions of light-years away.

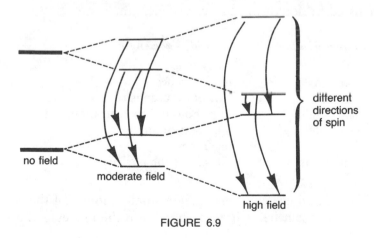

different
directions
of spin

no field

moderate field

high field

FIGURE 6.9

I should note in passing that this technique can also be used to measure the magnetic field in other inaccessible locations, for example, the inside of a complex molecule. Biochemists routinely use this sort of measurement to ascertain the environment at various points inside their molecules. It is not only magnetic fields in large and distant objects that we can measure, but those in very small, nearby ones as well.

The splitting of energy levels and spectra in a magnetic field is called the Zeeman effect. It is named after the Dutch physicist Pieter Zeeman (1865–1943), who received the Nobel Prize in 1902 for the discovery.

The Electromagnetic Spectrum

Visible light includes a very narrow range of wavelengths, as we have seen. This range is set by the sensitivity of the human eye, not by any law of nature. It is quite possible that an electron could make a quantum jump between energy levels and produce a photon whose wavelength is not visible to the eye. In fact, the waves shown in fig. 6.2, page 74, can be produced with wavelengths both much longer and much shorter than those associated with visible light. These types of waves are referred to collectively as "electromagnetic waves" or "electromagnetic radiation," and specific names are attached to waves of specific wavelengths. In the following table, the electromagnetic waves are listed in decreasing size of wavelength.

Name of Wave	Typical Wavelength
radio	many miles
microwave	feet to inches
infrared	tens of thousands of atoms
visible light	four thousand to eight thousand atoms
ultraviolet	hundreds of atoms
X ray	one atom
gamma ray	the size of the nucleus

The human eye can sense only a small segment of the wavelengths that can be emitted by objects in the universe. Stars in the skies emit all of the above types of radiation, from radio waves to gamma rays. Most of them travel the long light-years to earth only to be absorbed in the atmosphere before they reach the ground. Only visible light and radio waves can penetrate to telescopes at the surface of the earth, a fact that explains why optical and radio telescopes were the first major research instruments to be developed by astronomers. It is, in fact, only in this century that we have been able to put detectors above the atmosphere and "see" all the radiation that's there. This accounts for many of the remarkable discoveries of the past few decades, which have been a veritable golden age for astronomy.

The important point is that the same principles that apply to spectroscopy for visible light, particularly its ability to discern details of the composition and conditions of distant bodies, apply to every other kind of radiation as well. In modern times, the sort of analysis that Kirchhoff and Bunsen did for light has been carried out for every other type of electromagnetic radiation in the table. This gives astronomers a large number of arrows in their quiver as they attack the problem of understanding the nature of the universe.

Summary

Light is a wave, and different wavelengths correspond to different colors. Light is emitted and absorbed in bundles called photons. The more energy a photon has, the shorter its wavelength. Each color corresponds to a well-defined energy.

Electrons in atoms exist only in certain allowed orbits or energy levels. Photons are emitted when an electron falls from a high orbit to a low one, absorbed when the reverse takes place. This explains both the presence of discrete spectra of emitted light and the discrete dark lines associated with absorption. It is these lines that were measured by Fraunhofer.

Molecules, as well as atoms, produce spectra. Ionized atoms have slightly different spectra from normal atoms, a fact that allows astronomers to estimate the temperatures of distant objects. Various small effects within the atom cause the splitting of spectral lines, leading to what is called fine structure in spectra. This splitting also occurs when an atom is placed in a magnetic field—the so-called Zeeman effect. This allows astronomers to estimate the magnetic fields in distant sources.

Visible light is only one type of electromagnetic radiation. The radiation also comes in the form of radio, microwave, infrared, ultraviolet, X ray, and gamma rays. The principles of spectroscopy apply to all forms of radiation.

SEVEN

The Discovery of Helium:
A Crisis for Universality

And Lockyer, and Lockyer
Gets cockier and cockier
He thinks he's the owner
Of the solar corona.

—JAMES CLERK MAXWELL,
Scottish physicist

WITH THIS BIT OF doggerel, James Clerk Maxwell, the founder of the modern science of electromagnetism, gave his view of the character of one of the most interesting and influential of the Victorian astronomers. Not only was Sir Joseph Norman Lockyer (1836–1920) a well-known scientist and the founding editor of what is probably the most important general scientific journal in the world (*Nature*), he also played a major role in a drama that was every bit as crucial in the development of the idea of universality as was the recovery of Halley's comet: the discovery of the chemical element helium.

I never cease to be amazed by the ease with which people used to be able to enter the scientific professions. In our credentialed age, it is rare to find someone doing forefront research without a Ph.D. or its equivalent, and yet a century ago it was not at all unusual for amateurs to achieve positions of prominence and even leadership on the basis of raw talent and dedication alone. Norman Lockyer was just such a person. Like Herschel, he was born into a middle-class family (his father was a surgeon-apothecary in rural England), and, like Herschel, he first entered what would seem to be a scientifically unpromising career. In 1856, thanks to the intervention of an influential neighbor, he became a clerk in the War Office in London.

Lockyer was introduced to science by his father, and his post as a War Office clerk certainly gave him ample time to pursue his scientific interests. He wasn't expected to show up at his office until ten o'clock, and what with long lunches, an afternoon stroll in the park, and a four o'clock quitting time, he was not overly burdened with official duties. In the words of one writer, "there can have been few places in Victorian England which proffered greater time for self-improvement than the War Office." The overstaffing apparently came about because of the growth of the War Office during the Crimean War and the reluctance of political leaders to reduce the staff in the face of constant rumors of new wars.

Norman Lockyer's transition from government bureaucrat to astronomer, like other things in his life, took a rather unusual path. Early in his civil service career he was promoted and made officer responsible for putting the rules and regulations of the military services into some semblance of order and coherence. In this post he displayed two character traits that were often to reappear in his life: an enormous energy when confronted with a difficult task, and a less than tactful approach to opposition. He made many enemies in his new assignment, and when an economy drive provided an excuse, he was demoted and found his salary cut by two thirds. With a family and a home in suburban London, he had to seek outside sources of income.

One way to earn extra money was to write for the new general interest magazines that were starting to appear in England about this time. Because the Victorians valued things "scientific," and the educational system of the time included little science, there was a large audience available to anyone who could explain the new scientific advances in layman's terms. In addition to magazines and books, evening lecture series routinely drew standing-room-only audiences consisting of everyone from professionals to factory workers. Given Lockyer's interest in science, writing for this audience seemed to be a good way to deal with his financial problems.

He thrived in the scientific-literary milieu of London. There he made the acquaintance of Alexander Macmillan, whose publishing house still enjoys some standing in the world of letters, and Lockyer soon realized that next to all the new magazines there was an important gap—an empty publishing niche. Despite the enormous growth in science in the nineteenth century, there was no journal through which scientists could communicate quickly with each other. New scientific results were announced to the world in books or even in one of the public lectures mentioned above.

After an abortive attempt at bringing such a journal into print through the agency of a coalition of individual scientists, Lockyer convinced Macmillan to back a publication whose mission would be to

> First, place before the general public the grand results of Scientific Work. . . . Secondly, to aid Scientific Men themselves, by giving early information of all advances made in any branch of Natural Knowledge throughout the world.

Lockyer became the editor and organizer of the new journal, which was given the title *Nature*. Its first issue appeared on November 4, 1869, and Lockyer was to be its editor for a full fifty years.

It may well be that the launching and nurturing of *Nature* was Lockyer's most important contribution to science. *Nature* remains one of the premier scientific journals in the world and, in an age of increasing compartmentalization and specialization, one of the few places where important interdisciplinary work can be found. Having read and published in the journal for years, both as a working scientist and as a reviewer, I can attest to the fact that a century and a half after its inception, it is still going strong.

The foray that Lockyer made into the literary world had the effect of bringing him into contact with the leading scientists of the day. One by-product of this acquaintance was the purchase of some telescopes and the development of a new hobby—astronomy. Throughout the early and mid 1860s, he pursued this hobby along with his literary work. Where his clerkship at the War Office fit into his life at this time isn't known, but I doubt whether the level of his extracurricular activities were at all unusual. In one its periodic (and largely futile) attempts to get the War Office to tighten up its standards, a Civil Service commission was appointed in 1865 "to improve the efficiency of the [War Office] and get the clerks to understand that they are paid for work and not for literary distinction."

Lockyer chose an ideal time to launch a scientific career. The 1860s was not an easy time for astronomy. The introduction of spectroscopy had opened up entire new territories for exploration, and the Old Guard of the scientific community were more than a little upset about the rush of young people into this field. In the words of Admiral Smith, the author of one of the leading texts used by amateur astronomers at the time,

> With all my admiration for the marvelous power of chemistry in disintegrating the nature and properties of the elements of matter, I really trust that it will not be asserted among the Celestials [i.e., astronomers] to the disservice and detriment of the measuring agency; and this I hope for the absolute maintenance of GEOMETRY, DYNAMICS, and pure ASTRONOMY.

It must have been hard for men who had devoted their entire careers to careful measurements of the positions of stars to see people come into the field with no interest whatsoever in carrying on the old work. As so often happens in the sciences, the change occurred not because old problems had been solved, but because the attention of scientists was shifted to new problems. In a situation like this a young amateur, only loosely connected with the scientific establishment, was in an ideal position to exploit the new technology to the full.

Already equipped with a six-inch telescope in his backyard, Lockyer acquired a spectroscope and set out to explore the possibilities that Kirchhoff and Bunsen had opened up. He began by applying spectroscopic techniques to resolving an old question about the temperature of sunspots. By isolating light from a sunspot as it came through his spectroscope, he was able to show that the material in the sunspot was cooler than its surroundings.* This measurement established Lockyer as a major figure in British astronomy.

The next problem he tackled was a practical one. He wanted to find a way of studying the edge of the sun without having to wait for an eclipse. As the sun is the nearest star, it is the only one whose structure we can study in any detail. Part of this structure is the huge columns of flaming material that shoot up from the solar surface — the so-called prominences. Under normal conditions, the light from these prominences is completely blanketed by the light from the main body of the sun. Scientists who wished to study them had to go to considerable trouble and expense to set up laboratories to observe eclipses, when the rim of the sun was visible in all its splendor.

Lockyer wondered whether it might not be possible to see the prominences during a normal day with the aid of a spectroscope. His reasoning was that the instrument would spread out the diffuse background illumination over the whole spectrum from red to violet. This spreading out of the background should have the effect of making any spectral lines more visible. In October 1868, he used this technique to see a line associated with the element hydrogen in a prominence on the edge of the sun. His description of the event was graphic.

> I saw a bright line flash into my field [of view]. My eye was so fatigued at the time that I at first doubted its evidence, although, unconsciously, I exclaimed "At last!" . . . Leaving the telescope to be driven by the clock, I quitted the observatory to fetch my wife to endorse my observations.

Thus Lockyer became the first man to observe a solar prominence in broad daylight. His achievement was a major one, and eventually resulted in his election to the Royal Society. Yet it was not an achievement unaccompanied by controversy. In August of that same year, during an eclipse expedition to India, astronomers from many nations had observed the same sort of

---

* What he actually did was show that the Fraunhofer lines from the sunspot were darker there than in the sun in general. In modern terminology, we would say that this showed that the material in the sunspot absorbed more light, hence had more atoms with electrons in inner orbits, and was therefore colder than its surroundings.

spectral lines as Lockyer. Modern textbooks seem to divide as to which astronomer is to get the credit for the "first."

During this squabble, Lockyer continued to exploit his new technique. He had what modern experimental physicists call the "killer instinct" where new discoveries are concerned. Within a month of the sighting of the first spectral line, he had demonstrated that the prominences on the sun are part of an extensive outer atmosphere. He dubbed this atmosphere the "chromosphere" (a name still used by astronomers) because it was such a bright red color when seen in the spectral line of hydrogen he was using.

Disputes over priority of discovery were hardly new in the nineteenth century, nor are they unknown in our time. What is surprising about the Victorians, at least to the modern observer, is the extent to which the subject matter of a discovery was considered to be the property of the discoverer. It was considered bad form to "trespass" by making measurements of something that someone else had discovered without first obtaining that person's permission. Given Lockyer's aggressive personality and the conflicts over who was first to exploit the new spectroscopic techniques, it was inevitable that he should get embroiled in a series of squabbles over who owned the spectroscopic rights to the chromosphere. His behavior at this time is what led Maxwell to pen the little verse quoted at the head of the chapter.

Once the technique for seeing emission lines in the sun had been established, Lockyer quickly joined the pack of astronomers trying to discover the sun's chemical composition. In all the fevered work that was going on, the most interesting discovery from our point of view was the sighting of a few spectral lines that didn't seem to belong.

The element sodium—one of the constituents of ordinary table salt—gives off a yellow light when heated. It is this color that you see at freeway interchanges, where sodium vapor street lamps are used. If light from sodium is passed through a spectroscope, the yellow color is seen to arise from the presence of two bright and closely spaced lines, as shown in figure 7.1. The dark analogues of these lines were one of the most prominent features in Fraunhofer's solar spectrum, and he gave them a special name: He called them D lines, D_1 and D_2.

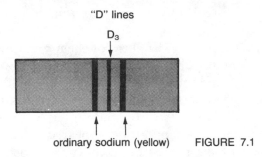

"D" lines

D_3

ordinary sodium (yellow)   FIGURE 7.1

What Lockyer and his colleagues discovered in 1868 was that in light from the sun there is a third line, midway between the two sodium lines. They called this new line D_3. Their instruments were sufficiently precise so that they could rule out any possibility that this new line was really something familiar that had somehow gotten garbled in the spectroscopes. It was without question a new and previously unseen spectral line.

To understand the impact of such a discovery at that time, you must remember that Kirchhoff and Bunsen were then identifying new chemical elements by discovering hitherto unknown lines in small samples of materials. As we saw, both cesium and rubidium were added to the list of chemical elements by these men's work. Another element, thallium, was identified on the basis of a single spectral line. Think back to the relationship between energy levels in atoms and the wavelength of emitted light and you will realize that claiming the discovery of a new element on the basis of the identification of a single spectral line is not necessarily a case of leaping to conclusions. If no known element produces that line, it means that no known atom has precisely that spacing between its allowed orbits. Since spectral lines can arise only from transitions between such orbits, it follows that there must be a new atom — one not yet seen in the laboratory — emitting the light.

When preliminary searches failed to find anything that could produce the D_3 line in the laboratory, Lockyer coined the name *helium* for the element associated with it. The name derives from *helios*, the Greek word for sun. In 1868 the name was appropriate, for evidence of this element existed only in the light from the sun — it had not been seen anywhere on the earth.

Thinking About Helium

In the early 1870s, there was a flurry of chemical elements being discovered, so the possibility that there might be one more in the sun hardly seemed to be earthshaking. It would only be a matter of time, people felt, until helium was isolated in our laboratories. But as time went by and no terrestrial sources of helium were found, this supposition appeared less and less tenable. For most of the rest of the century, scientists faced the possibility that there was indeed an element that existed in the sun and other stars, but did not exist on the earth. Such an outcome would have been disastrous for the principle of universality. If the stuff that makes up the stars cannot be brought into our laboratories that we may learn about its behavior, what hope have we of puzzling out how the stars work?

In the face of the continued refusal of helium to make its presence known on earth, the scientists of the nineteenth century adopted a number of different responses.

THERE'S NO HELIUM IN THE SUN, EITHER

At least at the beginning, there was some skepticism about the results that astrophysicists were reporting. The entire science of spectroscopy, after all, was only a few years old. It was a new and untested technology, and there was nothing to guarantee that the whole D_3 business wasn't just a mistake—an unanticipated flaw in the spectroscope.

Skepticism of new and puzzling experimental results is an almost universal response of scientists in all ages, but as time went by and more and better data came in, this position lost ground. With better and better data, the "no helium" argument shifted to another tack.

THE SPECTRAL LINES ARE EMITTED BY NORMAL ATOMS IN UNUSUAL ENVIRONMENTS

This was by far the majority view among people who thought about the problem. "It's all very well to talk about identifying an element like thallium from a single spectral line in the laboratory, but how can you be sure that the intense pressure and high temperatures in the sun don't affect atoms and change the characteristics of the light they give off?" Perhaps new and as yet unknown laws of nature operate in these regimes—laws different from those that operate here below.

Actually, there was some support for this view in data that Lockyer and his contemporaries had gathered. They found that the farther down they looked into the sun, the more smeared out the emission lines became. They went from being very sharp and narrow to being broad and diffuse. Apparently the increase in pressure in the sun's interior had an effect on the spectrum of emitted light.* So the idea that the extreme conditions on the sun could affect the emitted spectra of elements really wasn't so far fetched.

But broadening affects only the appearance of a line. It does not change the line's position in the spectrum. There is no reason to expect high temperatures and pressures to cause atoms to produce new colors, which is what those who held that the D_3 line was emitted by a normal atom would require. And in point of fact, the conditions on the sun aren't *that* extreme. The temperature at the surface is only six thousand degrees—a condition easily attainable in the laboratory. As time went on, experiments failed to turn up any evidence to support the notion that the lines seen by Lockyer and his colleagues could have been emitted by any of the atoms so far known.

This turn of events created a difficult situation. Once it was clear that reproducing the solar conditions in earthly laboratories failed to produce new colors from ordinary atoms, there was only one way to maintain the position that these atoms were responsible for the D_3 line. It became necessary to argue that light from the sun is different from light seen in our laboratories because the laws of nature themselves are different on the sun. Although few Victorian scientists were willing to make this statement explicitly, it was a logical consequence of the

---

* This so-called pressure broadening results from the increased number of collisions an atom suffers at high pressure. The atom is jostled in the process of emitting light, and the result is a small random shift in the wavelength of emitted light. These random shifts, added up over a large number of atoms, produce a wider spectral line.

statements they did make. Scientists who held this view were implicitly rejecting the principle of universality. ˙

HELIUM IS AN ELEMENT THAT EXISTS IN THE SUN BUT HAS NOT YET BEEN SEEN ON THE EARTH

This minority viewpoint has held by Lockyer, and championed with his characteristic energy and inattention to tact. He believed that helium would eventually be isolated as an element on the earth, and was not slow to point out that it was the chemist's responsibility to find it.

But there was another viewpoint possible for those who chose to accept this explanation of the solar spectrum. What if helium existed on the sun but *not* on the earth? This, too would violate the principle of universality, although in a different way than that associated with the majority viewpoint. Whereas proponents of the latter argued, in effect, that the laws of nature on the sun were different than those on the earth even though the chemical composition of the two is the same, the notion that helium exists on the sun but not on the earth implies that while the laws of nature in the two places may be the same, the chemical elements are not. Either way, the principle of universality is held to be incorrect.

These, then, were the responses to the presence of those mysterious lines in the sun. Unraveling the mystery of helium was complicated by the fact that it was not the only new element that was claimed to be seen in spectroscopes aimed at the skies. There was, for example, an element called "nebulium." This was seen in nebulae — large clouds of gas in interstellar space. Today we know that the light associated with "nebulium" are just spectral lines of ionized oxygen and nitrogen, but scientists at the time did not have the experimental capability to discover this. Then there was "Jargonium" — another element that was supposed to exist in the sun, but turned out to be an observational mistake. In this confusion of conflicting observational claims, is it any wonder that the answer to the challenge posed by helium could not be found immediately?

On the Other Side of the Decimal Point

I suppose I am as guilty as anyone of disparaging scientific work whose only goal is measuring the next decimal place. "It's just stamp collecting," I tell my students. Yet there is a frontier in high precision just as surely as there is a frontier at the edge of the visible universe. In both cases, small advances in technology can show you things you never suspected were there. And, in some cases, they can provide answers to questions that can be answered in no other way. This proved true in the helium controversy. It wasn't resolved by a brilliant theoretical insight or a breathtaking advance in observational astronomy, but by people who were working on measuring the next decimal place.

If you consult chemistry textbooks of the late nineteenth century, you will find the statement that the atmosphere is 21 percent oxygen and 79 percent nitrogen. These numbers were arrived at by a simple technique: Known substances like carbon dioxide and water vapor were removed from air, and then extremely precise chemical procedures were used to remove the oxygen from what was left and find its weight. When the percentage of oxygen was determined, it was subtracted from 100 to give the percentage of nitrogen.

This method will work well—provided that air is really made up only of oxygen and nitrogen. But it is not hard to see how uncritical acceptance of the "subtraction" method limited chemists in their search for substances, like helium, that exist in the atmosphere in small quantities.

It wasn't a chemist who finally forced scientists to recognize that there was more in the air than its two main elements. John William Strutt, Lord Rayleigh, is a man little known to the general public. He is, however, a "scientist's scientist"—a man who made advances in a large number of fields, from fluid mechanics to optics, and who was awarded the Nobel Prize in 1904 for his work in atmospheric chemistry. He was one of the last men capable of doing pioneering work in both theory and experiment. In 1892 he was engaged in an exercise in what I call "stamp collecting." He was trying to measure the weight of a liter of nitrogen to an accuracy never before attained.

What he found was puzzling. If he got his nitrogen in the conventional way by extracting the oxygen from dry air, that nitrogen weighed a little more than nitrogen obtained chemically—for example, by breaking down ammonia. The difference was only ½ of 1 percent (1 part in 200) but it was a real difference. Rayleigh tried to get rid of the discrepancy, but it didn't go away. He finally wrote a letter to *Nature* asking if anyone could suggest a reason for it.

He got an answer from Sir William Ramsay, a chemist at University College, London, who asked Rayleigh for permission to work on the problem (remember what I said earlier about the proprietary feeling Victorians had about scientific ideas?). Ramsay thought it likely that there was a hitherto undiscovered gas in the atmosphere, and proposed doing an experiment to isolate it. The two men put some clean air and oxygen into a container with water, then ran an electric spark through the gas. The spark caused the nitrogen in the air to combine with the oxygen and go into solution. When they had used up the two common gases, there was still a small bubble left—about 1/120 of the original volume. When the gas in this bubble was heated with a spark and the resulting light run through a spectroscope, a series of pale red lines were seen. In 1894, Ramsay and Rayleigh announced that a new gas had been discovered in the atmosphere. They named it *argon*, from the Greek word for lazy, or inert.

This discovery of a new component of the atmosphere led Ramsay to search for other sources of his new gas. There had been rumors floating around that some uranium containing minerals gave off strange gases, so in early 1895 he decided to see if that gas was argon. He took some of the mineral (called uraninite), ground it up, and subjected it to a series of chemical procedures designed to isolate the gas. Then he ran a spark through the gas and looked at the light it gave off through a spectroscope. At first, he couldn't believe what he was seeing and wiped the prism with his handkerchief. It didn't change anything: Against a background of a few faint red lines (emitted by trace quantities of argon) he saw a series of bright yellow and orange lines, unlike anything he had ever seen before.

After convincing himself that the yellow lines did not come

from sodium contaminating his electrodes, Ramsay sent samples of his gas to a number of colleagues for testing. On March 28, 1895, Lockyer got his sample. He quickly realized that it was not suitable for use in his apparatus, so he tried his hand at producing a new sample of gas from uraninite by a different method. He succeeded, as he tells in one of his books:

> From 30th March onward my assistants and myself had a very exciting time. One by one the unknown lines I had observed in the sun in 1868 were found to belong to the gas I was distilling from [uraninite]. Not only D_3 but . . . many other solar lines were all caught in a few weeks.

So the riddle was solved. The mysterious element whose spectral lines had been seen in light from solar prominences almost thirty years previously was found to exist on the earth as well. There was no reason to speculate about differences in natural law or chemical composition between the earth and the rest of the universe. Once again universality was found to be a reliable guide to the cosmos, and a potential stumbling block had been removed from the path of science.

And all this to-do about a gas whose main use today seems to be to inflate balloons at parties!

EIGHT

What Else Is in the Sun?

There are strange things done in the midnight sun.
—ROBERT SERVICE
"The Cremation of Sam McGee"

THE RESOLUTION OF THE helium problem changed the way the game of analyzing the chemical content of stars was played. Instead of asking, "What is there in the stars that isn't found on earth?" people started asking the opposite, "What is there on earth that isn't in the stars?" Throughout the early part of this century astrophysicists engaged in a mopping-up operation whose aim was to verify in detail the implications of the principle of universality.

Quite a few little questions had to be answered. By a stroke of coincidence, for example, the elements cesium and rubidium were not seen in the sun. These, you will recall, were the very first elements that Kirchhoff and Bunsen discovered with their new technique of spectroscopy. But the big question was theoretical: Assuming that the principle of universality is true, what do we expect to see when we look at the sun? This issue, as we shall see, was resolved through the work of a remarkable Indian physicist named Meghnad Saha. After the publication and verification of his work, no doubt remained that both the composition of matter and the laws of nature are the same here as they are anywhere else in the universe.

But old ideas die hard. Like the children of Israel returning to the golden idols, scientists keep going back to the old idea that the earth is different from the rest of the universe. The motive for this insistence is no longer philosophical, as it used to be. Today it grows out of an unwillingness to accept the verdict of experiment. For example, if you really believe something ought to exist but can't find it in terrestrial experiments, it is tempting to think that it may be "out there" somewhere. It is, after all, pleasant to believe that there are places where things are different—places where your ideas are right. Let me instance two modern heresies of this type, both from the late 1960s. One involves a claim that there are quarks in the sun, even if they can't be found on earth. The other involves the suggestion that the charge on the electron may have been different when light was emitted from distant galaxies from what it is here and now on the earth. Both these claims turned out to be wrong, but they illustrate the force that the notion of earth as a special place exerts on the human mind.

Understanding the Sun

The question of whether or not helium exists everywhere or only in the sun touched on things much deeper than stellar chemistry. Ramsay's discovery of helium in his uranium ore showed, to that extent, no separation between the earth and the heavens. But dramatic as this demonstration was, it was not a complete and final vindication of the principle of universality. To reach this goal, it was necessary to show that all the atoms we see in our lab exist in the sun and other stars, and that the laws that govern their behavior are the same in both places. Verifying these propositions is a complicated business.

Think for a moment about atoms in a hot gas like the sun. They move at high speeds, and when they collide, as they often do, electrons can be lifted to higher orbits or torn loose from their parent atoms. The photons emitted by the atoms add to the general melee. Atoms are constantly losing one or more of their electrons, regaining them, emitting energy in the form of photons, and reabsorbing it in a constant interplay. Add to this the fact that there are dozens of different kinds of atoms present in the sun, and you can see how complex is the system we wish to understand.

Actually, the problem of understanding the atoms in a gas is analogous to the problem of understanding a population of human beings. People are born and die, change occupations and place of residence. Yet in a stable population, every time someone is born someone else dies, someone gets a job when someone else loses one, and so on. Thus, even though individuals go through many changes, the general features of the population stay the same.

The same is true of collections of atoms. A given atom in the sun may start with its full complement of electrons, emit a photon, lose a couple of electrons, absorb and reemit photons, pick up an electron, and so on. But although individual atoms go through these changes, the number of that type of atoms in the gas with one electron gone will stay the same, as will the number with two or three or four.

In order to understand the light coming to us from the sun or another star, scientists need to know two things. First, what kind of light a given ion will emit—What, for example, is the spectrum of an iron atom that has lost three electrons? Second, they must know how many of each kind of ion there is, on the average, in the sun.

The first question was attacked by a number of scientists during the latter part of the nineteenth century. In fact, it could be characterized as *the* question being addressed by physicists during that period. There were a number of false starts—John Norman Lockyer, for example, championed a theory which implied that when an atom lost an electron, some of the lines disappeared from its spectrum, but the rest remained unchanged. As time went on, the current theory, outlined in chapter 6, was developed. Assuming that atoms in the sun behaved just like atoms in our laboratories, we knew exactly what to expect from each species in the sun. The only problem left was to sort out the shifting population of ions in the sun.

The solution came from an unexpected source. It was due, as I said earlier, to Meghnad Saha (1894–1956), an Indian born into a family of poor shopkeepers and educated with the help of local patrons and scholarships. From an early age he was involved in politics, and he was expelled from the equivalent of junior high school for participating in a demonstration for home rule. Nonetheless, he completed his education in India and assumed the position of lecturer in physics at the University College of Science.

Saha became interested in astronomy and began a program of reading a publication called the *Monthly Notices of the Royal Astronomical Society*. This is a journal in which professional astronomers record observations and short notes about their current ideas. The emphasis is on the quick dissemination of work in progress, rather than on complete exposition of fully thought-out theories. The task of learning astronomy from the *Notices* is roughly equivalent to that of learning economics by reading the stock market reports in *The New York Times*.

While doing this reading, Saha was teaching a graduate course in statistical mechanics — the branch of physics devoted to understanding the kind of shifting, changing populations of atoms described above. It isn't usually perceived that teaching this sort of course is an important learning experience for a young scientist. Nothing solidifies one's knowledge of a field or shows where the gaps in one's knowledge are like trying to to explain the subject to someone else. Experienced teachers do not exaggerate when they say that you never *really* understand a subject until you've taught it.

I remember an episode from my own days as a postdoctoral fellow at MIT that bears this out. Herman Feschbach, a senior physicist and the head of the theory group there, asked a group of us to prepare lectures on the research papers on which our work was based. At the time I resented this request, but it was one I could scarcely refuse. It took time away from my research, and I considered it something of an imposition as well as a waste of time. In the years that followed, however, I realized that those lectures may well have been the most important thing I had gleaned from that year. Over and over, I found myself explaining fine points to my colleagues—fine points that I understood because I had had to puzzle them out in preparing those lectures.*

What happened next to young Saha could not have occurred at one of the major European research centers of that time, where junior faculty members were closely watched and guided by senior scientists. By 1919, he had spent a couple of years learning about two seemingly unrelated fields—astronomy and statistical mechanics. In any European institution, he would have been advised to concentrate his attention on one or the other, for only by intense specialization could a young scientist, then or now, hope to advance. But Saha was at a provincial university, far from the centers of scientific thought. There was no one to tell him that he shouldn't think about bringing these two fields together to explain the solar spectrum. So he tried it, and produced a theory that is still used by astronomers to deal with light from the stars. The outcome of this theory is a prediction of the spectrum we expect to be emitted by the sun, based on the assumption that atoms in the sun are not different in any way from atoms on the earth.

You can understand how the theory works by looking at figure 8.1. To start things off, suppose we consider the element calcium in the state in which three electrons have been removed. This is called Ca IV (the numbering system starts with Ca I, the state with all electrons present). Calculations of the speed of calcium atoms in the sun and the frequency with which they collide leads to a prediction of how many ions like Ca IV ions will be present in the sun. This, in turn, tells you how intense the spectral lines associated with this ion will be. The precise location of those lines can be obtained by looking at Ca IV in the laboratory. This process is repeated for every other state of every other element—we show a few representative cases in the figure. The total spectrum from the sun can then be obtained by superimposing all of these spectra on each other, as shown for the representatives in the figure.

---

* While writing these lines, I recall that I have never properly acknowledged my debt for this part of my training. Thanks, Herman.

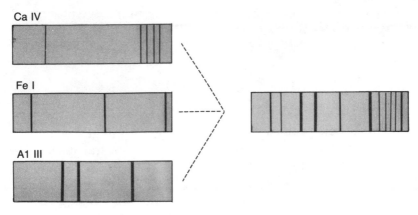

FIGURE 8.1

The spectrum we actually measure when we look at the sun, then, is quite complex. It contains many thousands of spectral lines. Yet it can be analyzed by a process of subtraction. One by one the sets of lines associated with each of the hundreds of possible ions are identified and removed from the spectrum until at the end there is nothing left. At this point the spectrum has been completely explained.

This process of the final unraveling of the puzzle of light from the sun began with the publication of Saha's paper in 1920. A number of other small matters were resolved at the same time. For example, the absence of evidence for rubidium and cesium that we mentioned above was seen to be a consequence of the details of the atomic structure of these elements. The theory predicted that the lines of these elements would be very faint —too faint to have been seen up to that point.* Light from these elements was in fact seen shortly thereafter, during the period when the predictions of the Saha theory were being tested and verified by observation.

Today, the use of the sophisticated descendants of the Saha theory has allowed astronomers to understand the details of the spectra of the sun and all the other stars we can see, even in the most powerful telescopes. It is fair to say that we have

---

* The expert will realize that this is a consequence of the unusually high ionization potential of these atoms.

proved that the principle of universality as applied to atoms and light has been demonstrated out to the limits of the observable universe.

Quarks in the Ultraviolet

But when new problems arose, the faith of some researchers in the idea that the earth is typical of the rest of the universe began to waver. In the mid-1960s, one of the great new adventures in physics was the search for quarks. These hypothetical particles, whose existence has still not been unambiguously demonstrated in the laboratory, are supposed to be one of the basic building blocks of matter. I won't take the time to explain why scientists suggested quarks might exist or the role that they continue to play in our theories. Anyone interested in this subject can find it explained in some detail in my book *From Atoms to Quarks* (Scribners 1979). For our purpose we simply note that quarks are thought to differ from ordinary particles in that their electrical charges are either one-third or two-thirds of the charge on the electron, the quarks coming in both positively and negatively charged varieties.

One way you could imagine confirming the existence of quarks would be to find evidence for "quarked atoms." These hypothetical objects are atoms in which a positive quark is incorporated into the nucleus of an ordinary atom or a negative quark replaces an electron in orbit. Since the forces that determine the location of the allowed orbits in any atom are the electrical attractions and repulsions between the various particles in the atom, the replacing of any particle by a quark (whose charge is different from that of the particle it replaces) will change the location of all the orbits. This, in turn, means that the spectrum of a quarked atom will be different from the spectrum of an ordinary atom. It should, in principle, be as easy to detect quarked atoms (should they exist) as it was for Kirchhoff and Bunsen to detect cesium and rubidium. In particular, if there are quarked atoms in the sun and we follow the kind of subtraction procedure outlined above, there should be spectral lines "left over" at the end—lines that do not belong to any known atom.

By 1966, it was becoming clear that if quarks existed on the earth, they weren't going to be easy to find. The first round of searches were coming up empty, and people were starting to worry. If the quark is really the fundamental constituent of matter, it should be plentiful in nature. It was in the context of this initial failure of the search for quarks that the following episode occurred.

A group from Yale University and the Naval Research Laboratory was examining some of the first data on the ultraviolet part of the solar spectrum. This data had been obtained from telescopes flown above the atmosphere. Ultraviolet light—light whose wavelength is shorter than visible violet light—is absorbed by the atmosphere and hence had not been studied extensively up to that point. After going through the subtraction procedure for the ultraviolet part of the spectrum, the group found a few lines left over. The most prominent of these had a wavelength of about 1,690 atomic radii. The group was very cautious in making a positive claim that this was evidence for a quarked atom: They spoke of the "most probable quark line" in their paper. Nevertheless, excitement about their paper ran high among the community of elementary particle physicists.

So here once again there was a situation in which there seemed to be a breakdown of universality. As had happened almost a century earlier with helium, there appeared to be things in the sun that couldn't be found on earth. This supposed breakdown of universality was more serious than the one involving helium, however, because it came at a time when scientists felt they really understood the sun. There was a lot less room to maneuver in 1966 than there had been in Lockyer's time, fewer unknown processes that could give rise to the observed spectral lines. So for a short time, universality was once again called into question.

But the uncertainty didn't last long. A week after the claim for a possible quark line appeared in print, W. R. Bennet, Jr., also of Yale, sent off a paper that resolved the entire problem. Using a computer code that allowed him to calculate the expected spectral lines from atoms that had lost two, three, and four electrons, he was able to show that all of the lines that had been listed as "unexplained" by the astronomers could, in fact, be attributed to emissions from nitrogen, oxygen, and carbon in various stages of ionization. The "most probable" quark line, for example, was attributed by Bennet to an emission from the carbon atom with one missing electron.

So in short order, another assault on universality was put down. The outcome was different from what it had been for helium, however. Instead of finding quarks on the earth, it was shown that the "quarks" in the sun were simply things we already knew on our home planet, but hadn't recognized.

Had it not been for the fact that this little episode took place about the time I, as an impressionable graduate student, was getting interested in the quark model, I probably would have forgotten it long ago. It would have been one more episode, one more wrong idea corrected and set aside. Many such are buried in the scientific journals, waiting to teach us lessons about the working of the scientific enterprise.

I must admit that the lesson I took away from this story is rather more personal. It is disconcerting indeed to discover that the scientific papers that were "hot stuff" in the early stages of one's career are now valuable primarily as history. Has time really gone by that fast?

The Electrical Force: Constant or Growing?

Though it is hardly publicized, there is a streak of something close to mysticism in most theoretical physicists. Our next example deals with a result of this fact.

Back in 1937 the British Nobel laureate P. A. M. Dirac noticed a curious fact. The ratio between the electrical and gravitational forces that two protons exert on each other, regardless of how far apart they are, is a very large number—about 10^{38}. (This is a one followed by thirty-eight zeroes.) The ratio between the time it takes light to cross the universe and the time it takes light to cross an electron is also very large—about 10^{37}. Dirac looked at these two numbers and argued that it could hardly be a coincidence that they are so close. He inferred that there must be some as yet unknown theory that would predict that they should be equal.

Since the universe is expanding, the time required for light to cross it will increase as the universe gets older. Therefore, in order for the equality of the two numbers to be maintained for all time the ratio of electrical to gravitational forces would also have to increase in time. Relying on this argument, Dirac suggested that the gravitational force early in the history of the universe was stronger than it is today, and that is has been diminishing ever since the start of the universe.

You are probably not overwhelmed with the power of this argument. I have to admit that I'm not either. In fact, had it not been that this reasoning was put forward by a man who had a record of making seemingly outrageous suggestions that turned out to be right, I doubt if the paper would have been published or taken seriously. It smacked too much of numerology. Nevertheless, a number of scientists did take the time to point out that if Dirac were right, then the sun would have been much hotter in the past than it is today. Not only would the earth's oceans have boiled, but the higher rate of burning would have extinguished the sun's fires long ago. In the face of these criticisms, the Dirac hypothesis was quietly shelved.

It was revived in 1967, however, when the Russian-American theoretical physicist George Gamow pointed out that there was another way to salvage the Dirac hypothesis. Instead of letting the gravitational force get smaller, why not let the electrical force get bigger? The effect would be the same: The ratio of the electrical to the gravitational force would grow as the universe expanded. In Gamow's words,

> It would be too bad to abandon an idea as attractive as Dirac's proposal. And one unwillingly asks oneself is it not possible that, while [gravity] remains constant, [the electrical force] increases in direct proportion to the age of the universe.

The status of this proposal with respect to the principle of universality is somewhat ambiguous. My reading of the motives of the people involved is that they were not saying the laws of nature were different in the past from what they are now. Rather, they were using the coincidence of the two numbers as a clue to discovering the true laws of nature. Even so, in the absence of a reason other than the rather thin numerical coincidence for supposing the law of electrical attraction differed in the past, it seems to me that Gamow and Dirac's arguments were suggesting something with the same ad hoc flavor of all the other anti-universality ideas we've been discussing. And, as we shall see, they met a similar fate.

For it didn't take long for the physics and astronomy communities to respond to Gamow's paper. Several authors pointed out that if the electrical charge had been weaker in the past than it is now, certain atomic nuclei would have been radioactive and would not exist today. The most interesting argument came from two astronomers at the California Institute of Technology, John Bahcall and Maarten Schmidt. Their refutation of the Dirac-Gamow hypothesis rested on an analysis of the fine structure in spectral lines (see chapter 6).

The key point in their work was the fact that we know of many galaxies billions of light-years away from the earth. The light we are receiving from them now was emitted from atoms in those galaxies billions of years ago. If the electrical force was really different then, evidence for that difference should be carried in that light like a fossil footprint in a rock.

One's first thought is that one should be able to tell if the electrical force has changed by looking at whether the wavelength of a familiar spectral line is the same in light from a distant galaxy as it is in laboratories on earth. Unfortunately, this won't work because (as we shall see in chapter 14), light from distant galaxies is shifted toward the red end of the spectrum by other effects. But it turns out that the fine structure in spectral lines carries information enabling us to say whether the electrical force was different at the point where the light was emitted. Figure 8.2 gives the reason we can make this statement. On the left, we show two orbits in an atom, the upper being split as described in chapter 6. This part of the diagram shows the situation for a normal electrical force — one that is the same everywhere in the universe.

Suppose we assume that the electrical charge was weaker in the past than it is now. What would happen to these energy levels? The answer is illustrated, in slightly exaggerated form, on the right. All the orbits would move farther out, as you would expect, since the force holding the electrons to the nucleus is less than it was before. The crucial point is that the splitting

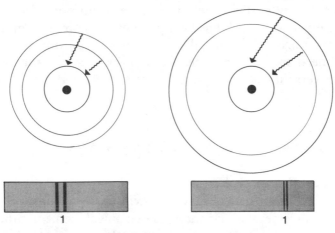

FIGURE 8.2

between the levels in the upper orbit — the splitting that gives rise to the fine structure in the spectrum — decreases much more.* The spectra emitted by the atoms in the two cases are shown below the diagrams in the figure. The key point is that a disproportionate change in the splitting between the two fine structure lines — the quantity labeled 1 in the figure — is what signals a change in the electrical force. This is very different from the normal red shift, which shifts all lines proportionately.

The upshot is that a change in the electrical force could be detected by measuring the fine structure in spectra from distant sources. If the disproportionate change is not there, then the electrical force must have been the same when the light was emitted as it is here and now on the earth. Bahcall and Schmidt looked at a group of radio galaxies — galaxies that emit large signals in the radio part of the electromagnetic spectrum. All the galaxies were billions of light-years away and their signals consequently were emitted billions of years ago. If the electrical force were indeed smaller in the past, evidence for it should certainly be found in light from these objects.

When the data were analyzed, no evidence for any change in the electrical force was found. In fact, Bahcall and Schmidt were

---

* To be precise, the splitting varies as the square of the strength of the electrical force, while the overall level shift varies as the first power.

able to show that the electrical force could not have changed by so much as 0.00000000001 percent per year: Any more than this would have produced a shift in the splitting that would have been seen in their telescopes. So once again, the principle of universality is borne out, in this case to thirteen decimal places!

A Summing Up

So the work of Joseph Fraunhofer in his remote Bavarian monastery has borne fruit. From our own star to the distant reaches of the universe, we find the same chemical elements, the same laws of nature that we discover in our laboratories. Had it turned out otherwise, the enterprise we call science would scarcely have been possible. But the long path of experiment, observation, and theory has led us to a profound insight into the nature of the world in which we live.

But, as the question of the constancy of the electrical force shows, there is another frontier for universality. Even if we grant that the laws of nature operate everywhere in the universe at the present time, how can we be sure that these same laws have operated at all times in the past? This temporal frontier has proved more difficult to cross than the spatial frontier we have been discussing up to this point. Nor is the temporal battle completely won even now, as the public debate over creationism shows.

It is to this second frontier, then, that we will now turn our attention. In order to do so, it will be necessary to go back to the eighteenth century, when a physician by the name of James Hutton was contemplating a particular formation of rocks near the town of Jedburgh in his native Scotland.

NINE

The Discovery
of Deep Time

*If the stone . . . which fell today were to rise again to-
morrow, there would be an end to natural philosophy
. . . and we could no longer investigate the rules of nature
from our observations.*

—JAMES HUTTON

GREAT SCIENTIFIC insights have never been achieved all
at once by one person acting alone. Our deepest ideas
about the universe do not change suddenly, but over
long periods as new edifices are built by succeeding generations
of scholars. This is the way the world is, but it's not the way
we want the world to be. The human mind seems to demand
heroes, in science as in other fields of endeavor, and if history
is not so obliging as to provide us with heroes to fit our needs,
we go ahead and create the heroes anyway.

To illustrate this point, here are two heroic tales from the
annals of the conventional history of science. The first concerns
Nicolaus Copernicus, the second James Hutton.

Copernicus, the story goes, was a Polish monk who isolated
himself from his fellows, choosing to spend the better part of
forty years in a stone tower scanning the night sky. In the end

he put together our modern picture of the solar system, in which the sun was at the center and the planets move. This view immediately revolutionized our notions of the earth's place in the universe, sweeping away forever the superstitious rubbish of the Middle Ages.

Hutton's legend is similar. A Scottish farmer and physician, he wandered from place to place looking at rocks. He observed what he saw closely, never framing hypotheses until he had incontrovertible evidence to support them. In the end, he developed our modern view of the geological process, proving that the old ideas about the age of the earth and the role of catastrophes like Noah's Flood were wrong.

Both of these are beautiful stories, answering a deep need in the human psyche. Unfortunately, neither has much relation either to the lives of the men involved or to the contributions they made to the progress of science.

Take Copernicus. He was not a monk, but a canon in a major cathedral in northern Poland. As such he was responsible for managing the considerable estates of the church. He was very much a man of affairs, serving on several royal commissions and having a status somewhat analogous to that of the mayor of a large city or the governor of a small state in modern America. Astronomy was merely a hobby for him, not an all-consuming occupation.

Furthermore, although he did fit out a small tower on the cathedral grounds as an observatory, he made very few measurements of the heavens himself. His theory was based almost entirely on observations that had been made by Greek astronomers a millennium and a half earlier. The model of the universe he published did indeed have the sun at the center and the earth in orbit around it. Except for this, however, it was not much of an improvement over the old Greek theories that had dominated medieval thought.

Nor did his ideas immediately sweep through the European scientific community. For one thing, his book *On the Revolution of the Spheres* is written in such a turgid style that only the most dedicated reader is likely to be able to work through it. Consequently, Copernicus needed a spokesman—someone to explain his writing to others. For Copernicus, this spokesman was

Galileo, who popularized the heliocentric system in *Dialogue Concerning the Two World Systems*. It was only much later that the work of many generations of astronomers turned the Copernican insight into our modern understanding of the solar system.

Why, then, do we honor Copernicus today? We honor him for two reasons. First, he had the courage—the intellectual daring—to think a new thought: Despite the evidence of the senses it was the earth and not the sun that moved.* Second, and more important, once he had conceived this thought he had the fortitude to devote forty years to working it out—to make it something more than a passing fancy. In the end, we honor Copernicus because he showed us that there was an intellectually respectable alternative to the ideas about the universe that were the accepted wisdom of his time.

In many ways, James Hutton (1726–1797) played a similar role among those eighteenth century scientists who thought about the earth and how it came to be the way it is. Reading about his life, I am repeatedly struck by the parallels to the life of Copernicus. Both men thought new thoughts and made them intellectually respectable. Both were, to put it bluntly, terrible writers. And both have become what Stephen Jay Gould calls "cardboard figures" in an official iconography that obscures the very real contributions they made and the very human way they made them.

Hutton was born in Edinburgh, the only son of a wealthy merchant and city official. Although his father died when he was three, Hutton was left enough money to be free to follow his own desires, without ever having to worry about earning a living. He had what I would consider to be an unpromising start in his intellectual life, studying at Edinburgh University, trying out a legal career as an apprentice, quitting, entering medical school in Edinburgh but never graduating, going to Paris for a couple of years, then on to Leiden, in Holland, where he finally took his M.D. in 1749.

All in all, this is not a résumé that one would expect from a

---

* A few obscure Greek astronomers had suggested this idea, but none produced a workable model of the solar system based on it.

man who is now thought of as one of the great scientists of all time. In fact, it bears more than a passing resemblance to what one might find for one of today's perpetual students—young people who enjoy the university life so much that they never want to leave. They can be found forming a penumbra around most major universities, supporting themselves any way they can while they change from one field of study to another.

This impression of Hutton is reinforced by his postuniversity career. He never practised medicine, but after a brief (and highly successful) fling at running a manufacturing concern and more travel, he decided that what he really wanted to do was to be a farmer. In addition to his money, his father had left him a small farm in Scotland and Hutton resolved to turn it into a model establishment by using all the new farming techniques being developed around Europe.

The early 1750s found him traveling around Britain and Europe, visiting farms and studying agriculture. It was apparently during this period that he acquired his interest in geology. It would be almost impossible to visit farms in different regions and not be aware of the differences in soil. In addition, we know that during this period he often combined his agricultural visits with trips to local points of geological interest.

From 1754 to 1768—a period of fourteen years—Hutton lived a quiet life on his farm, building it up into the state of the art establishment he wanted. When this was done, he apparently lost interest in the enterprise, leased out the farm, and moved to Edinburgh.

It would not be an understatement to say that during this period Edinburgh was the intellectual capital of Britain, and perhaps of all Europe. Among Hutton's friends and dinner companions were Joseph Black, one of the founders of modern chemistry, James Watt, the inventor of the modern steam engine, and Adam Smith, the founder of the modern science of economics and author of *The Wealth of Nations*. Hutton continued his study of geology while at Edinburgh, making frequent trips around England and Wales. It was, I suspect, these frequent field trips that formed the basis for Hutton's reputation as a keen observer of nature. No armchair theorist would have visited so many outcroppings, seen for himself so much of the evidence of the earth's antiquity. Finally, in 1785, after three decades of intense interest in geology, Hutton announced his conclusions—conclusions that he would eventually elaborate in detail in his book *The Theory of the Earth*.

Hutton's Theory

To understand why the ideas I am about to discuss are so important, you have to understand something of the notions concerning the earth that prevailed in the late eighteenth century. This was a time when educated people, insofar as they thought about the problem at all, assumed that the earth had been recently formed, more or less as described in Genesis. For them, the pivotal event in earth history was Noah's flood, and much effort was expended in explaining the present state of the earth in terms of that single cataclysmic event. When Hutton's contemporaries thought about ultimate causes—when they asked how the earth got to be here in the first place—they tended to fall back on the notion that at the beginning of the earth the processes at work in nature had been different from those we see today. Even so formidable a personage as Isaac Newton, when contemplating the problem of the creation, was led to comment:

> When natural causes are at hand God uses them as instruments in his works, but I do not think them alone sufficient for [explaining] the Creation.

By now, you should be familiar enough with the principle of universality to recognize the situation among earth scientists in the eighteenth century. They thought that there was a set of laws that operated to shape the surface of the earth today—laws whose nature can be discerned by careful observation. They also thought that these laws had not operated at the beginning of all time. The principles behind Noah's flood and Newton's Creation are, by their nature, unknowable and inaccessible to human observation and reason. Even Isaac Newton, the originator of the idea of universality in its modern form, was strangely reluctant to extend his principle to the creation of the earth.

117

Hutton's thinking on this matter was guided by his extensive travels and his experience in agriculture. He knew that erosion continuously carried soil down to the sea, and he argued that given enough time it would eventually wear down even the highest mountains. Like many scientists of his time, Hutton was a deeply religious man, and the notion that God would have created the world only to tear it down again was profoundly repugnant to him, as it was to many of his contemporaries. An eternal earth was particularly important to him because this planet is, after all, the abode of life. To make the earth eternal, he needed to find a mechanism to counteract the effects of erosion.

The theory that Hutton finally presented to the Royal Society of Edinburgh in 1785 was profoundly unlike anything that had gone before it. He proposed that the earth had not simply been created at some time in the past and allowed to run down, but that it operated on a continuous cycle, in which land was destroyed by erosion and was created anew by geological processes. Furthermore, he argued that the same processes that we see operating on the surface of our planet today are sufficient to explain the entire history of the earth. Unlike Newton, Hutton was not willing to entertain the possibility that different laws acted in the past. In the language we have been developing in this book, the final implication of Hutton's theory is that the principle of universality applies to time as well as space. The laws of nature we discover here and now are the same laws that acted to shape the earth in the past.

One simple example will, I think, make the general outline of Hutton's ideas clear. As a farmer, he was aware that soil is constantly being eroded, and as a geologist he knew that the soil that washed away from his farm eventually found its way to a river delta, where it was deposited as sediment on top of other soils. This process of sedimentation can be seen all around the world, the Mississippi Delta being a familiar example. The layered sediments eventually turn into sedimentary rock, and this rock carries the record of its genesis in its appearance. You may have seen sedimentary rocks in road cuts—the effect is something like looking at the pages of a book end on. Each "page" in the rock bears witness to the layering process that formed it.

Hutton's problem was twofold. First, he had to find a way for the eroded soil to be recycled—to be brought back up to dry land. Otherwise, the process by which rocks break down into soil and soil washes away to the sea would eventually destroy even the highest mountains. If there was not some way to counteract this process, the land would eventually be worn away. This, of course, would violate Hutton's belief that the earth was created as a home for living things. Second, he had to explain why it is that sedimentary rocks are often found in mountains: How do they come to be located thousands of feet in the air when they are created below sea level?*

The solution to both these problems is found in Hutton's notion of uplift. The idea is quite simple: After sedimentary rocks are created, there are geological processes that lift them up so that they can be eroded again. This cycle of erosion, recreation, uplift, and erosion would, as far as Hutton could see, go on forever, giving us an earth that is both eternal and ever changing.

Evidence for the view that the geological history of the earth contains cycles of uplift and erosion is all around us. In figure 9.1 I sketch one such piece of evidence that Hutton used—a formation of sedimentary rocks near Jedburgh, Scotland. You see two layers of such rocks, one tilted and overlaid by the other. Since both layers must have formed horizontally on the ocean floor, the only way that the outcropping at Jedburgh could have come into existence is as shown in the figure. The lower layer must have formed as shown, then been lifted above ground and tilted, allowing the top was worn smooth by erosion. After that the rock was submerged again, either because the sea rose or the land subsided. The second layer was then laid down in the usual way. This simple explanation of what geologists call an "unconformity" was what Hutton proposed and is the one we would give today.

---

* The conventional wisdom at the time was that mountain sediments were deposited during Noah's flood.

119

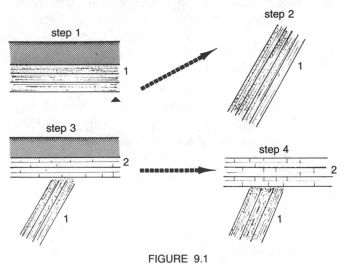

FIGURE 9.1

But the most important consequence of Hutton's view of a formation like the one at Jedburgh is that the processes involved require enormous amounts of time. If the earth is not just running down—if there is a process of renewal—then the earth must be old beyond imagining. John McPhee called this the discovery of "Deep Time," which is, I think, a beautiful way of putting it. We will devote the entire next chapter to a discussion of the reaction to this aspect of Hutton's work. We note in passing, however, that once Hutton developed the notion of the geological cycle, he held that the cycle had been in operation forever, and hence that the earth was eternal.

The Proof of the Pudding

When the new theory of the earth was read to an attentive audience in Edinburgh in 1785, it was merely an interesting hypothesis. Like any hypothesis, it needed to be tested. Simply stating that the earth ought to have a continuous existence was not enough. By Hutton's time the practice of "armchair science"—cooking up good ideas about the way the earth could be explained without trying to test those ideas against observation—had fallen into bad repute. In order to make his ideas respectable, Hutton had to buttress them with observations. And although what he said was plausible for sedimentary rock, a great difficulty arose when he considered rock such as granite.

We now know that granite rock is formed by heat, like lava. Such rock came to the surface of the earth in a molten state and hardened into the rocks we see now. We call them igneous (fire-formed) to call attention to their origin. In Hutton's time, however, these rocks were thought to have been formed at the Creation and to have existed unchanged through all time. There seemed to be no alternative explanation, no scientifically reasonable way to form hard, crystalline rocks like granite from a sedimentary cycle. Explaining granite, then, became the main test of the Huttonian system of the earth.

It was clear that if Hutton was going to make his theory of

eternal cycles work, he was going to have to deal with the presence of granite on the earth's surface. If granite really was a primordial rock—if it really had been present at the beginning—then all the granite should have been worn away long ago. Its persistence could only mean that the earth was young, as Hutton's opponents claimed. In that case, Deep Time would be a myth. If, on the other hand, granite was formed by processes similar to those that can be observed forming lava today, the difficulty vanishes. The granite we see would be just one more reincarnation of the eroded granites of times past.

Hutton assembled evidence for the continued creation of granite from two sources: examinations of the rock itself; and looking at the way the rock was placed with respect to other rocks in the field. Most granite is in the form of ordinary crystalline rock, but occasionally it comes in an interesting form. A friend sent Hutton a piece of granite from an outcropping at Portsoy, in northern Scotland. Viewed under a magnifying glass, the rock appeared to have a series of partially melted quartz needles embedded in it, as shown in figure 9.2.

Hutton realized, rightly, that this particular piece of granite carried within it a clue to its origin. In his words,

> It is not possible to conceive any other way in which these two substances, quartz and [the background rock], could be thus

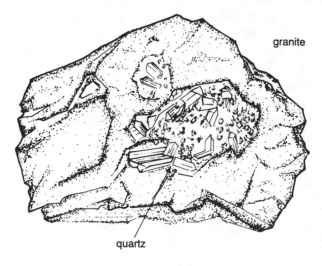

granite

FIGURE 9.2

quartz

concreted, except by congelation [cooling] from the liquid state, in which they had been mixed.

This evidence, published in his original paper on the theory of the earth, was apparently enough to give him the courage to move ahead. It showed that it was not impossible that the hard, crystalline granite rocks he saw at the surface of the earth might have been in a quite different state in the past. This fact, in and of itself, was not enough to prove his thesis, but it did indicate that even granite might be part of the type of evolutionary cycle he proposed for other rocks.

The discovery of the smoking pistol for Hutton came a few months after he had read his paper to the Royal Society. He thought he might be able to find evidence for his theories in the mountains near Aberdeen and was invited by the duke of Athol to join a hunting party along the Tilt River in the summer of 1785. The action of the river had exposed a set of mixed rocks along the walls of its glen. In the words of Hutton's biographer, John Playfair,

> When they reached the forest lodge, . . . Dr. Hutton found himself in the midst of the objects which he wished to examine. In the bed of the river, many veins of red granite were seen traversing the [local sedimentary rocks]. . . . The sight of objects which verified at once so many important conclusions, filled him with delight. . . . the guides who accompanied him were convinced that it must be nothing less than the discovery of a vein of silver or gold, that could call forth such strong marks of joy and exaltation.

The formation that caused Hutton such joy is sketched in figure 9.3. The veins of granite are insinuated between the layers of limestone (a common sedimentary rock). Looking at the figure, it is obvious that the limestone must have been there first, and the granite must have flowed into the cracks while in a molten state. There is simply no other way to explain the formation. The inescapable conclusion was that the granite had formed in a molten state *after* the limestone was in place, and therefore could not have been present since the beginning of the earth. Thus, Glen Tilt (and other formations like it that were discovered later) provide conclusive evidence that even granite

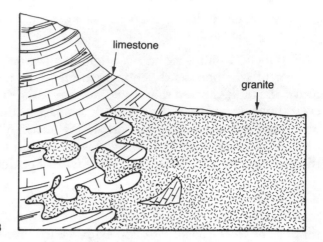

FIGURE 9.3

must go through a cycle in which it is created, broken down, and created again.

Of course, scientists at the time had no idea of how such a process could occur or what laws of physics would govern the transformation of the rock.* Hutton talked about heat and pressure in the earth's interior, and at one point suggested that the expansion of the rocks on heating provided the force needed for uplift, but it is fair to say that he never really got this whole business right. What is important is that the existence of places like Glen Tilt shows that these transformations *must* occur — that granite must be created, destroyed, and created anew. And if the chemists and minerologists can't figure out how it happened, so much the worse for them.

Was Hutton "Scientific"?

Hutton died a few years after the publication of his monumental two-volume *Theory of the Earth: With Proofs and Illustrations* in 1795. Like Copernicus, who saw a first edition of his book on the day he died, Hutton did not live to see the reaction to his work. And just as Copernicus had his Galileo to explain his

---

* A more complete description of what geologists call the "rock cycle" is given in my book *Meditations at 10,000 Feet* (Scribners, 1986).

ideas to the educated public, so Hutton had his friend John Playfair, professor of mathematics (later natural philosophy) at Edinburgh University, to perform a similar service for him.

In 1802, Playfair published a book called *Illustrations of the Huttonian Theory of the Earth*. The purpose, according to the "advertisement" in the front of the book, was to explain

> Dr Hutton's theory in a manner more popular and more perspicuous than is done in his own writings. The obscurity of these had been often complained of; and thence, no doubt, it has arisen ... that so little attention has been paid to the ingenious and original speculations which they contain.

It is in Playfair's writings that the notion of Hutton as the supreme inductivist and field geologist was born, and it is to Playfair that most modern geology texts owe their characterization of Hutton's work. From what we have seen of the genesis of *The Theory of the Earth*, it is clear that a true understanding of Hutton is going to be much more complex than this image. There was clearly more than simple observation involved, and Hutton's ideas about the way the world ought to be played a major role in the development of the theory.

The discovery by modern historians of science that Hutton did not come to his theory by pure observation has led to a new kind of cardboard legend — one in which Hutton is seen as just one more armchair philosopher whose claim to fame is that he happened to be more right than wrong. Some historians point to the fact that Hutton didn't see Glen Tilt until after he had formulated his theory as a proof that he was not fully "scientific." It seems to me that both tradition and countertradition are wrong. When I look at the way Hutton went about putting his theory together, I see a man who behaved almost exactly the same way as any good modern scientist would behave.

Consider the way that Hutton came to his theory in the first place. Even though his papers have been lost, so one cannot be entirely sure, there is little doubt that theological considerations constituted an important part of the attraction of the theory in his mind. It may even be, as some have suggested, that the origin of the theory was entirely due to Hutton's need to have a benevolent purpose guiding the earth. Does that make the theory "unscientific"? Of course not.

For one thing, although we cannot trace the details of the origin of the theory, we do know that when he announced it in 1785 he had decades of field work behind him. If he had that much field experience and *hadn't* formulated a theory, he would be unlike any geologist I've ever known. The idea that all this experience was ignored in order to support his religious feelings doesn't seem reasonable to me. I suspect that while the idea was germinating, Hutton used his field work to guide him, just as a modern theoretical scientist uses experimental results for the same purpose. Certainly, his treatment of the evidence provided by the Portsoy granites bears this out. Hutton clearly used them as indicators to guide arguments that were not fully fleshed out until later.

Finally, as I pointed out in chapter 3, there are two steps in the elaboration of a scientific theory—proposal and testing. From from the point of view of the finished product, *it makes no difference* where the original idea comes from. Theories can (and have) originated in dreams, in seemingly inexplicable bursts of insight, and as the final product of long, grueling work. What matters is not where the idea comes from, but how it is treated after its birth.

Furthermore, as we have seen, the production of a theory did not end Hutton's interest in the problem of the evolution of the earth. He continued to gather data to confirm his ideas, as the episode at Glen Tilt shows. Thus, Hutton did what any modern scientist would do. On the basis of partial evidence and his own ideas about the world he formulated a hypothesis, and then gathered data to support it.

If that's not "scientific," I don't know what is!

Charles Lyell and Uniformitarianism

If the writing of Playfair made Hutton's work accessible to a wider audience, it was Sir Charles Lyell who put together the most comprehensive theory using the ideas that Hutton had first stated.* Although Lyell, like Hutton, was born in Scotland, his family moved to southern England when he was one year old and he enjoyed an upbringing typical of the upper-class English of his time. While at Oxford, he was introduced to geology, and spent some of his summers traveling around looking at various formations. He studied law, and became a barrister in 1822.

Like Hutton, Lyell was a man of independent means, and some biographers have suggested that he practiced law because it gave him an excuse to get away from the social whirl in London. In any case, he traveled extensively throughout his life, visiting many places of geological importance. In the early 1830s he put his reading and field work together into a book that was to have an enormous influence on the course of science, the *Principles of Geology*. This multivolume work was nothing less than a survey of all the processes that shape the face of the earth, with copious illustrations from sites around the world. It was the first comprehensive work of its type, and seeing it through twelve editions became Lyell's lifetime occupation. It was also an extremely influential book — Charles Darwin carried it with him on the *Beagle*, for example. In it Lyell puts forward a view of the earth's history that has come to be called "uniformitarianism" (although the term did not originate with him).

The basic idea of uniformitarianism is that all features of the present earth can be explained in terms of the accumulated effects of processes that we can see and measure here and now. Lyell's books are full of arguments and examples to support this point of view. On a trip to Sicily, for example, he measured the accumulated lava from eruptions of Mount Etna to show that entire mountains could have been built in slow stages. On one of his trips to America he estimated the rate at which Niagara Falls is receding toward Lake Erie and argued that its present position could be explained in terms of the slow wearing away of the rocks by the Niagara River.

To understand the importance of Lyell's grand vision of the

* Lyell lived from 1797 to 1875, and was a major figure in Victorian science. The name is pronounced with the accent on the last syllable.

earth in our discussion, we have to make a distinction between three different concepts—a distinction that Lyell himself was not always careful to draw.

The laws of nature that operate today operated in the past. This is what I have called the principle of universality, except that up to this point we have been concerned only with verifying that the principle works throughout the universe—with extending it through space. With Hutton and Lyell, we begin to see the prospect of extending the principle on a time axis as well.

The same processes operate on the earth today as operated in the past. This is the definition of uniformitarianism normally found in textbooks, where it is usually stated in terms of the motto "the present is key to the past." Lyell definitely espoused this point of view, arguing against the idea that there have been processes in the past (like Noah's flood) which cannot be seen operating today. The basic idea is that the processes that shape the earth are always the same, no matter what period of time we are discussing.

Processes in the past operated at the same rate as they do today. This idea, usually called "gradualism" today, is what Lyell defended in his books. It says that the workings of the earth were never much different in the past from what they are today, and that there have never been any sudden catastrophes or discontinuous changes on our planet.

All these statements express the general notion that the present moment in history is not markedly different from moments in the past. All of them assert that it is possible to do science by assuming that current conclusions hold at earlier times. They differ markedly, however, in the restrictiveness with which this notion is applied.

For example, universality would imply that Newton's Law of Gravitation must have operated in the past much as it does today. But it would make no distinction between the gravity causing a rock to roll down a hill (a process we see all around us) and the same force causing a massive asteroid to strike the earth (an event that would be catastrophic in every sense of the word). Both uniformitarianism and gradualism would accept the first of these events, but deny the second. The impact of a large meteorite would be seen as a singular and special event, and could not be accommodated in either point of view.

By the same token, it is possible that the process of uplift could be episodic—that it occurs sporadically and is not always visible on the earth. A uniformitarian would have no problem with this idea, but a gradualist would. In this sense, we would say that Hutton was a uniformitarian while Lyell, the "father" of that doctrine, was actually a gradualist. Hutton would have been content to say that there may have been times in the past when uplift and vulcanism were not creating new surface area, provided that these processes operated often enough to keep his cycle going. Lyell, on the other hand, would have argued that this could not be the case—that all features of the present earth can be explained in terms of the accumulated action of processes that we can see and measure here and now.

Over the course of time, scientists have come to realize that the history of the earth is probably characterized by occasional singular and even catastrophic events like meteorite impacts set against a background of slow, steady changes of the type described by Hutton and Lyell. But the influence of Lyell has not died among his scientific heirs. A slight aura of heresy still clings to any but the most uniformitarian doctrines in the earth sciences.

With the work of Lyell, then, we find the development of the principle of universality following a new line. Our attention is shifted from the cosmos back to our own planet, and shifted from extending the principle in space to extending it backward in time. The question becomes not, How far out can we apply the laws of physics? but, How far back can we go and still be confident that these laws apply? And before long, the geologists who were extending their domain backward in time ran into an unexpected but formidable set of enemies—the physicists.

TEN

The Age of the Earth

*And if there was no birth time of earth and heaven and
they have been from everlasting, why before the Theban
war and the destruction of Troy have not poets as well
sung other themes? Whither have so many dreams of
men so often passed away, why live they nowhere embodied in the lasting records of fame?*

— LUCRETIUS
De rerum natura

THE IDEA OF A cyclic history for the earth, as proposed by
Hutton and Lyell, is a very attractive one from the point
of view of the geologist. It certainly goes a long way
toward explaining the features we see today on the surface of
the earth. From a philosophical point of view, it runs into opposition from two camps: those who, for various reasons, want
to believe the world is young; and those who insist on asking
questions about the time when the cycles started.

The opposition from the Young Earthers is easy to understand. If Hutton was right—if Deep Time really exists—then
the earth must have been created long before humanity and
recorded history came on the scene. The most beautiful statement of a man's dismay over this prospect is the passage from
Lucretius quoted above. We can call this the argument from

poetry: How could it be that time had gone by without being recorded by the great poets? I have to admit that I feel a certain resonance with this point of view, although I know it to be wrong.

A more contemporary young-earth viewpoint is associated with Biblical literalists, who believe that the account of creation in the Bible must be taken at its face value and an age of some thousands of years assigned to our planet. We'll discuss the details of this argument below. For the moment we simply note that the idea of Deep Time has always been opposed by some religious groups and that these groups were quite powerful in nineteenth century England, where much of the debate about the history of the earth went on.

More surprising is the opposition Hutton and Lyell's ideas encountered from scientists, particularly physicists. The problem here arose because of a curious omission in the work of Hutton and Lyell. Both of them argued forcefully for the sort of cyclical processes we described in the last chapter, but neither of them ever gave much thought to how (or when) the cycles started. Insofar as they gave the matter any attention, they seemed to think that the earth and its cycles had simply gone on forever. In fact, Hutton finished his original paper on *The Theory of the Earth* by saying:

> If the succession of worlds is established in the system of nature, it is vain to look for anything higher in the origin of the earth. The result, therefore, of our present inquiry is that we find no vestige of a beginning—no prospect of an end.

A word about Hutton's archaic language: by the "succession of worlds" he meant the cycles of uplift and erosion, and by "anything higher" he meant a cause or event that would start the cycles off. For Hutton, at least, questions about the age of the earth were meaningless. It has always been as it is now.

The problem was that during the nineteenth century physicists were developing the branch of science we now call thermodynamics. They were learning that there is no free lunch in the universe. If energy is expended at one point, it must be generated somewhere else. They were also learning that there is no such thing as a perpetual motion machine, and in looking at Hutton's earth they saw something that looked suspiciously like one. The cyclical earth seemed to violate their newly won insights into the laws of nature.

The physicists' problem was that Hutton and Lyell seemed to have overshot the mark. The earth might indeed be more than a few thousand years old, but the laws of physics would not accomodate an earth that was eternal. If we are to take the principle of universality seriously and assume that the laws of nature operated in the past as they do today, then we have to assume that *all* the laws, even the new laws of thermodynamics, have been involved in shaping the earth's history. This seems to be an eminently logical point of view, but when the physicists applied it to the history of the earth, they came up with limits on the planet's age that seemed to the geologists uncomfortably short.

So the proponents of the new geology found themselves embroiled in controversy with two different camps, one holding that the earth is young, the other holding that, though it is old, it is not eternal. In these battles, the concept of universality was sharpened and forged into something closely resembling its modern form.

The Young Earth

The most famous name associated with the Young Earth school is that of Archbishop James Ussher, a seventeenth-century Anglo-Irish clergyman and scholar. During a life that was fully involved in the religious and political wars of his times, particularly in the problems associated with reconciling the English and Irish churches, he found time to become a scholar of the Old Testament. He doubtless regarded his calculation of the age of the earth as a mere footnote to a career that included being a bishop and a role as adviser to King Charles I, but it is because of his calculations that his name is known today.

His method was simple, at least in principle. To find the date of creation, all one need do is go back through the Bible, adding up life spans as one goes. The total of those life spans will then be the age of the earth. This procedure is not as easy as it sounds, but Ussher carried it through. His conclusion: The earth was created in 4004 B.C.—about six thousand years ago. Later on his methods were refined by Dr. John Lightfoot of Cambridge University to give the time of creation as Sunday, October 23, 4004 B.C., at nine o'clock in the morning. This datum was duly recorded in the concordance of the King James Bible. Although it is not part of the Bible proper, its appearance in all editions of the King James Version until the early twentieth century gave it a stature close to that of revealed truth.

It is, of course, easy to make fun of Ussher today. When I was in graduate school, for example, geology students still threw an all night party on October 23 to celebrate the "foundation" of their field of study. I find this attitude uncharitable, if understandable. Ussher was a man trying to cope with a very difficult problem with the best tools at his disposal. After having tried to duplicate his calculation (about which more anon), I am inclined to have a lot more respect for him than is the norm these days.

The fact of the matter is that reconstructing the chronology of the Bible is no easy task. True, in Genesis we find long strings of verses like:

> And Enos lived ninety years, and begat Cainan: And Enos lived after he begat Cainan eight hundred and fifteen years, and begat sons and daughters. . . .
> And Cainan lived seventy years and begat Mahalaleel. (Gen. 5:9–12)

On the face of it, it would seem to be a simple matter to add up these generation spans and reproduce Ussher's calculation. Unfortunately, the well-defined chronology of the kind just quoted runs only up to the prophet Abraham. After Abraham and his son Isaac, the names of the generations and the "begats" are chronicled, but the life spans and times involved are not. Ussher and his contemporaries developed a complex system of symbolism to interpret various events in the Old Testament, attaching dates to them and allowing the scholars to extend the "begats" to the birth of Christ.

When I discovered this fact, I was rather discouraged. One of the little rewards I had promised myself when I started writing this book was a good look at Ussher's chronology. I've always been curious about his methods. But the prospect of spending weeks slogging through old theology books wasn't what I'd had in mind.

Fortunately, about this time I ran into a young friend of mine, a geologist at the University of Chicago named Douglas Macayal. Doug spends a lot of his time measuring the flow of glacial ice on the Antarctic ice cap. Indeed, I have a vivid memory of a lunchtime story about his being caught in a plane in Antarctica during a whiteout. In a whiteout, clouds and snow turn the sky the same color as the ice pack, and you can't tell where the ground is. Apparently experienced pilots have learned to deal with this condition by putting their heads out the window of the plane and guiding their landings by listening to the sound of the engines bouncing off the ice below!

In any case, Doug had wanted to do the Ussher calculation for his class of freshman science students. Running into the problem I mentioned, he devised a way to carry out Ussher's calculation in an approximate way. The basic idea is this: He splits the Biblical generations into two groups: the first (up to Isaac), for which exact durations are given; the second (to Jesus), for which they are not. He then calculates the average generation time for the period from the flood to Isaac and assumes that this average applies to the second period as well. Simple multiplication then gives you an estimate of the age of the earth. His calculation is summarized in the table on page 135.

From Jacob (the son of Isaac) to Joseph (the father of Jesus) there are thirty-eight generations. This gives a date for the creation of 4006 B.C. And although this method doesn't give the sort of precision that geology students commemorate in their parties, it captures the spirit of the calculations that led to the notion of the Young Earth.

It should be emphasized that the great majority of modern

GROUP I

Event	Time
Creation	7 days
Adam	130 years
Seth	105 years
Enos	90 years
Cainan	70 years
Mahalaleel	65 years
Jared	162 years
Enoch	65 years
Methuselah	187 years
Lamech	182 years
Noah	500 years
Shem	100 years
Arphaxad	35 years
Salah	30 years
Eber	34 years
Peleg	30 years
Reu	32 years
Serug	30 years
Nahor	29 years
Terah	70 years
Abraham	100 years
Isaac	60 years

Average generation time since Noah = 50 years

religious leaders do not take the kind of arguments given above very seriously.*

My reading of the situation obtained in the early nineteenth century is that few scientists took the Young Earth seriously. There were some, like the chemist Richard Kirwan, who op-

---

* If you would like to know more about this subject, I recommend a book titled *Is God a Creationist? The Religious Case Against Creation Science*, edited by Roland Frye (Scribners, 1983). In that book a variety of religious leaders (including Pope John Paul II) argue that the kind of strict literalism implicit in Ussher's approach is in no way a necessary concomitant of religious faith.

posed the new ideas on what were essentially religious grounds. He said he had come to oppose Hutton and Lyell's arguments

> by observing how fatal the suspicion of the high antiquity of the earth has been to the credit of Mosaic history, and consequently to religion and morality.

But this was definitely a minority view among scientists and much of the laity. Even the most ardent catastrophists eventually began to argue that the earth had been around for a long time before Noah's flood. But by the middle of the century, Lyellians had something more substantial to worry about—an attack coming from the most prominent physicist of their generation.

Lord Kelvin and the Age of the Earth

William Thomson, who later earned the title of Lord Kelvin, was born in Belfast in 1824, the fourth child of a mathematician and textbook writer. When he was eight, the family moved to Glasgow, which was to be his home for the rest of his life. He was a prodigy, entering Glasgow University at the age of ten, publishing his first scientific papers at sixteen, and, after a brief stint at Cambridge, taking the Chair in Natural Philosophy at Glasgow at twenty-two.

His subsequent career established him as one of the leading and most visible scientists of his time. He not only made important and lasting contributions to physics, but he was a key player in many of the important technological projects of the nineteenth century. He served, for example, as chief scientific consultant for the company that laid the first transatlantic telegraph cable. Not only had he worked out the theory for the propagation of signals over long wires (no mean feat, considering that the modern theory of electromagnetism was only then being developed), but he invented and patented numerous devices (including the telegraph receiver) that were incorporated into undersea cables everywhere in the world. He even sailed on the ships that laid the first cables. As a result of these services to Britain, Thomson was first knighted, then elevated to the peerage as Baron Kelvin by Queen Victoria.

His genius for practical things, coupled with his deep insight into physical processes made him a prominent figure in Victorian England — something of a cross between Thomas Edison and Albert Einstein. The royalties from his patents also made him rich, and at one point in his life he acquired something we ordinarily don't associate with physicists — a 126-ton ocean-going yacht.

My own introduction to the practical side of Kelvin's genius came in an unexpected way. In the 1970s, during the height of the back-to-the-land movement, I acquired an abandoned farm in rural Virginia and proceeded to build a house on it. While I was laying brick for the fireplace, I was faced with the problem of learning how to put the interior together (it's not as simple as it sounds). Being a scientist, I wasn't content merely to find a recipe, I wanted to understand *why* fireplaces are built as they are. Imagine my surprise when I learned that the modern fireplace was designed (and probably patented) by none other than Lord Kelvin. His key addition was the smokeshelf (see fig. 10.1), an addition that prevents downdrafts in the chimney from blowing smoke into the room. There was many a stormy night in

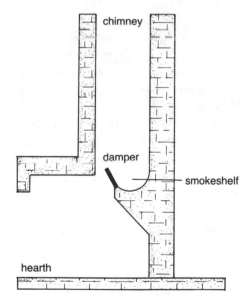

FIGURE 10.1

my house when I, with a crackling fire in front of me, felt that Kelvin had richly deserved his yacht.

His greatest contribution to science was not the fireplace, but the formulation of what is now called the First Law of Thermodynamics. Published in its modern form in 1851 in a paper entitled "On the Dynamical Theory of Heat," the law is also known as the principle of conservation of energy. What it asserts is simple to visualize. Energy—the ability to exert a force over a distance—is a quantity that can neither be created nor destroyed. In any system, energy can change forms, but the total amount can never change. You can, for example, use energy from your muscles to lift something up, then let it fall, converting the original chemical energy in the muscles to energy of motion of the falling body. What you cannot do, according to Kelvin, is have more energy in the fall than you had in the muscles at the beginning. The books of nature have to balance at all times.

Kelvin realized that heat could be used to balance the books along with other forms of energy. When you rub your hands together to warm them, you are making use, unconsciously, of Kelvin's insight. The energy in the motion of your hands is converted to heat through the action of friction. The energy of motion came from the chemical energy in your muscles, which came from the food you ate, and that energy came (ultimately) from sunlight.

The vision of the universe that Kelvin gave us, then, is one in which energy flows in ceaseless cycles. Whenever it appears in one form, it must have been converted from some other form. And, as a physicist, whenever he saw energy, he always asked the question, Where did it come from?*

Contemplating the endless cycles of the earth's history being put forward by Lyell and Hutton, Kelvin saw a system that was expending energy. It would, after all, take a lot of energy to lift an entire mountain chain from the sea bottom to ten thousand feet. But although the energy was being expended, there seemed to be no place for it to originate. In Kelvin's view, Hutton's *Theory of the Earth* was nothing more than an elaborate plea

---

* Kelvin was also involved in formulating the Second Law of Thermodynamics, which embodies the concept of entropy.

for a free lunch. In 1863 he put his objections into concrete form in a paper titled "On the Secular [i.e., long-term] Cooling of the Earth," published in the *Philosophical Magazine*, a journal of the Royal Society. The paper opens with the following statement:

> For eighteen years it has pressed on my mind, that essential principles of thermo-dynamics have been overlooked by those geologists who . . . maintain not only that we have examples before us, on earth, of all the different actions by which its crust has been modified in geological history, but that these actions have never . . . been more violent in the past than they are at present.

In the paper that follows these remarks, Kelvin works out a simple problem — one that could be done by any graduate student in physics today, although it was a state-of-the art calculation in 1863. Reading the paper carefully, one can find in it a number of assumptions, some stated and some implicit. They are:

> 1. The energy to run the geological cycle comes from heat in the earth's interior.
> 2. There is no way in which this heat can be replenished in the earth, so we are, as it were, living on a diminishing stock of energy "capital."
> 3. For purposes of the calculation, the earth can be treated as a solid, uniform, undifferentiated sphere.

Miners had long known that when you penetrate deep into the earth, you encounter regions where the temperature increases. This is a fact of life for modern mining engineers as well. A planner at the Homestake Gold Mine in Lead, South Dakota, once told me that when they open new shafts ten thousand feet below the surface, they expect the rock temperature to exceed 130 degrees Fahrenheit! If the temperature increases with depth, heat must be flowing out from the surface of the earth, from the warm interior to the cooler surface.

Given his assumptions, Kelvin interpreted this outward heat flow as evidence that the earth was cooling off. Applying some fairly sophisticated mathematics to the problem of reconstructing past surface temperatures, he found that given the present rate of cooling, the surface of the earth would have been just cool enough for rocks to solidify about 100 million years ago. He acknowledged that there were some uncertainties in his numbers, and argued that while 200 million years was improbable, 1,000 million years was "impossible." Later, as we shall see, he was to become considerably less flexible in his pronouncements.*

With the publication of "Secular Cooling of the Earth," the battle was well and truly joined. On the one side were Kelvin and the physicists, the pioneers of the new wave in science. They would not allow the world to be over 100 million years old. On the other side were the field geologists (and soon the biologists) who knew in their bones that it would have taken longer than that to produce all the formations like the one at Jedburgh and to evolve the marvelous complexity of the planet's biosphere.

Clearly, both sides couldn't be right. What is especially interesting to me about this confrontation is that both sides were advancing their positions under the banner of universality. It wasn't a question of *whether* the laws of nature were to be extended into the past, but of *which* laws were to be so honored. Were we to believe the physicists, with their general principles and arcane mathematics, or the geologists, with their rocks and dreams of cycles? For fully a half century after the publication of Kelvin's challenge to the geologists, the question remained unanswered.

Reactions to Kelvin

Whatever else can be said about the long-standing quarrel over the age of the earth, it was not an ivory tower debate of interest only to academics. The question received a great deal of attention in the popular press, particularly because it clearly bore on the validity of the Darwinian theory of evolution. Even Mark Twain found occasion in his *Letters from the Earth* to comment:

---

* For reference, the dinosaurs disappeared 65 million years ago, and came into existence about 225 million years ago.

Some of the great scientists ... have arrived at the conviction that the world is prodigiously old, and they may be right, but Lord Kelvin is not of their opinion. ... In order to be on the safe side ... he believes that it is [100 million years] old, but not older.

The reception of Kelvin's work by the scientific community was mixed, as one might expect. In general, physicists tended to support him, geologists and biologists to be critical. I shall return to the debate in more detail in the next chapter. For the moment, however, I want to concentrate on one important outcome of "On the Secular Cooling of the Earth." Whatever else it did, however many feathers it ruffled, it forced geologists to start thinking about their subject in a quantitative way. It also forced them to move away from the indefinite or infinite life span for the earth that had been sufficient for Lyell and Hutton and to think seriously about the possibility that our planet must have been born at a definite time in the past.

By the early part of the twentieth century, in fact, geologists had devised a number of ways to estimate the age of the earth. These methods, together with the estimates of age they give, are listed below.

THICKNESS OF SEDIMENTS

This method involves measuring the total thickness of sedimentary rocks of different periods from around the world, adding them up to give a number representing the total thickness of sedimentary rock created since the beginning of the earth. Some estimate of the rate of creation is then made by looking at present sedimentation, and the amount of time needed to create the total observed thickness of sedimentary rocks is calculated. In 1915, geologists estimated the thickness of the entire sedimentary column as 335,000 feet (64 miles) and the rate of deposition as between 2 and 6 inches per century. This leads to an estimated age of the earth in the neighborhood of 100–130 million years for the time needed to create the column.

MASS OF THE SEDIMENTS

This method is similar in principle to the previous one. The idea is that the total mass of sedimentary rocks on the earth is estimated, rather than their thickness. This mass is then compared with the total mass of material being carried to the sea by the world's rivers. Assuming that the present rate of sedimentation is typical of the past, this gives an estimate of roughly 47–188 million years.

THE SALTINESS OF THE SEA

If one assumes that the ocean was originally fresh water, and that runoff from the land has been making it progressively more salty, one can estimate the age of the earth by dividing the total amount of salt in the oceans by the amount carried in by the rivers each year. This yields an estimate for the age of about 80 million years.

I should interject for the reader who may be unfamiliar with modern geological chronologies that none of these methods would be considered a satisfactory way to measure the age of the earth today because they all ignore the cyclical nature of geological processes. Salt, for example, does not simply reside in the ocean, but is removed by various chemical processes and replenished. Similarly, sedimentary rocks go through a continuous cycle of creation and destruction. Thus, the ages given above represent the time it takes for one cycle (chemical or geological) to be completed, but give no indication of how long the cycles have been going on.

In any case, by the end of the nineteenth century geologists were starting to come to grips with the challenge posed by Kelvin and the Laws of Thermodynamics. It seemed that they could live with an earth that was 100 million years old. The problem was that by that time, Kelvin had revised his numbers downward, and was claiming that the the earth might be only 20 million to 40 million years old.

There was just no pleasing the man!

ELEVEN

Models, Reality, and the Arrogance of Theoretical Physicists

It is, however, an understood thing that scientists of a particularly elevated kind — theoretical physicists, for instance — may from time to time express quietly authoritative opinions on the conduct of scientific inquiry, while the rest of us listen in respectful silence.

—Sir Peter Medawar
Pluto's Republic

Theoretical Physics

KELVIN'S GENERATION saw the creation of a new type of scientist. Before the mid and late nineteenth century, it was assumed that a scientist would interest himself in both the theory and the practice of his field. Isaac Newton, for example, worked out a great deal of the theory of optics and investigated the philosophical consequences of his ideas. At the same time, he did numerous experiments on light, including the famous demonstration that sunlight could be broken up into all the colors of the rainbow and then reassembled into a beam of white light.

By the end of the nineteenth century, physicists were starting to establish a division of labor. Both mathematical and laboratory techniques were becoming more complex, and it was becoming difficult for one person to master both. For all of his practical expertise, Kelvin was able to make most of his contributions to science because of his mathematical ability. From an early age, he had mastered the new mathematics of his age.* At the same time, some of his contemporaries were devoting themselves to developing the increasingly complex electrical apparatus that was being introduced into the laboratory. Gradually, a split was forming in what had been a single science. Theoretical and experimental physics were being born as separate specialties of study.

Today, this split is incorporated into our educational system. At some point during the second year of graduate school a student makes a choice. He or she will specialize either in theory or in experiment. You can't do both, and you can't delay the choice. Like the man who became Lord Jim in Conrad's novel, you are forced to make a single decision that will determine the entire future course of your life. Even the choice of a marriage partner is not so final as the choice of specialization. Few people ever come back later in life to try the career path they rejected in their youth.

The existence of the specialization known as theoretical physics has created an unfortunate kind of stratification in the sciences — a stratification that Medawar refers to in the quote at the start of this chapter. In an eerie echo of the old Platonic disdain for those who work with their hands, there is a pecking order in modern sciences by which those whose work is more abstract and farther removed from physical reality are accorded the highest status. That Sir Peter Medawar, a Nobel laureate in Medicine and Physiology, should feel compelled to make a sarcastic comment about this state of affairs is, I think, a testimony to the strength of the feeling about it in the scientific community.

I like to illustrate the status accorded to abstraction in modern science by talking about what I call the Great Academic Chain of Being. It goes like this:

Engineers want to be experimental physicists.
Experimental physicists want to be theoretical physicists.

* For the expert, I'll mention that this was what is now called the field of partial differential equations.

Theoretical physicists want to be mathematicians.
Mathematicians want to be philosophers.
Philosophers want to be theologians.

And, of course, to close the whole thing off, we can propose a final entry:

Theologians want to be engineers.

It is impossible to understand the response to Kelvin's claims about the age of the earth without appreciating this distinction between theoretical and experimental scientists. Some of the hostility he generated among geologists was, of course, the result of a simple turf war—the resentment of an outsider. But a lot of it also had to do with the fact that Kelvin was the quintessential theoretician, arrogantly pointing out the error of their ways to the field geologists, who were, as Medawar says, expected to "listen in respectful silence." But before we talk about the way the geologists reacted to Kelvin, it seems appropriate to digress for a moment and discuss exactly what it is that a theoretical physicist does.

Models and Reality

The task of the theoretical physicist is to construct a mathematical model of the real world. This is somewhat analagous to the job of translating poetry from one language to another. It is possible to make such translations, and some translations are better than others, but something always gets lost in the process. The question one asks of the work of the translator and of the theoretical physicist is whether what is lost is an unimportant detail or the heart of the original.

In physics, the language into which we translate the phenomena of nature is mathematics. This is a rigid, precise language, with little room or tolerance for ambiguities. It is a language with well-defined rules of "grammar," so that once a statement is made in the language, the consequences of the statement follow inevitably. Learning the language of mathematics occupies a great deal of the training of a theoretical physicist, and operating in the language can be difficult and time consuming. Nevertheless, the real problem faced by a theoretical physicist is not the manipulation of the mathematics. The central problem—the task that raises the discipline to a creative art form—is converting the complexity of the natural world into clean, precise mathematical statements.

A simple example will show what I mean. Think about a ball rolling down a hill. This is a simple process, one that we would expect any first year science or engineering student to deal with. The motion of the ball is described by Newton's Laws of Motion, and once the equations associated with these laws are written down, they can be solved easily.

But let's think about the transition from a real hill to the mathematical equations in more detail. A real hill might be covered with grass or dirt. It might have leaves or branches lying on it, perhaps even some bugs. It will surely have some irregularities, little bumps and valleys. By the same token, a real ball will have a rough surface. It might have grooves in it, like a soccer ball, or stitches, like a baseball. It will almost certainly be slightly out of round. Such a hill and ball are sketched on the left in figure 11.1.

A theoretical physicist looking at a real ball rolling down a real hill immediately starts to pare away these complexities. A very different picture of the ball and the hill is created in his or her mind. A "physicist's eye view" of the ball rolling down a hill is shown on the right in the figure. The real hill is replaced by a perfectly straight incline; the real ball with a perfect sphere.

real hill   physicist's hill

FIGURE 11.1

The complexities of nature have been discarded in favor of a version of the problem that is clean, pure, and simple.

Clearly, something is lost in this "translation." The grass, the branches, the bugs — everything, in fact, that would make the hill a pleasant place to rest on a summer afternoon — are all gone. No one would want to lie on a theoretical physicist's hill and watch the clouds drift by. But we gain something to compensate for this loss. In return for giving up some of the realities of the situation, we get a clearly defined, easily manipulable problem. And, if the job has been done correctly, the "translation" is close enough to the original to allow us to use it to describe and predict what happens in the real world.

The real art of the theoretical physicist, then, is to know which part of the complexities of nature can be discarded and which must be retained. For the case of a ball rolling down a hill, for example, we have assumed that the grass and the bumps are inessential details and that the general incline is the most important feature of the hill, as roundness is for the ball. Under these assumptions, we replaced the hill by the smooth slope on the right of figure 11.1.

Were these assumptions correct? There is no way to answer this question within the mathematical system. As we pointed out in chapter 3, we have to move to the second stage of the scientific process. In this case, that means that we have to roll some balls down hills and see if the results match the predictions. If they do, then the assumptions were probably justified and we can conclude that we have indeed captured the essence of the relation between the ball and the hill. If the results don't jibe, we go back to the drawing board and try again with a different model of what is and isn't important about the hill. Eventually this process of trial and error produces a theory of hills that covers all of the experiments we can do. At that point we say that we have produced a theory of the rolling of a ball down a hill.

There are many ways that our replacement of the hill by a smooth incline could fail. If the surface of the hill is made of sand (so that the ball might get bogged down), or if there is a large hole in it, we clearly would not get the right predictions from our "translation." A theoretical physicist must therefore always be aware that his or her work rests on assumptions about what is and isn't important in any situation, and that those assumptions can be wrong.

You would suppose that this sense of fallibility would breed a tentativeness, not to say humility, in the profession. Sometimes it does, but more often the high status accorded to those who deal with mathematical abstractions intervenes, blinding the physicist to the feet of clay that support every theoretical construct. It is this inability to see the possibility of error that gives some theoretical physicists the supercilious attitude that makes them so infuriating to their colleagues. For all his eminence, I'm afraid that Lord Kelvin provides us with an almost perfect illustration of the arrogance of the theoretical physicist by his behavior during the debate on the age of the earth.*

The Reaction to Kelvin's Age for the Earth

With this caveat in mind, we can now appreciate that Kelvin's calculation of the age of the earth was but a "translation" of the real earth into a simplified theoretical model. Kelvin's earth, like the smooth incline, was an approximation to a complex reality. In his case, the key assumptions were: (1) that there are no energy sources inside the earth; and (2) that the earth is rigid and of uniform composition throughout. Given these assumptions, he could make a mathematical model of the earth that was simple enough to be solved, and based on that solution could estimate the age of the earth.

Had Kelvin done his calculation today, critics would immediately have asked two questions: How do you know those assumptions are true? and How sensitive are your conclusions about the age of the earth to changes in your assumptions? In

---

* You may think I am being unduly harsh on theoretical physicists. This may be so— they may be no worse on average than other scientists. Yet as a theoretical physicist, I want to hold myself and my colleagues to the highest possible standards. We may not always measure up, but we should always remember what those standards are and be candid about our failings.

fact, I doubt if the paper could even be published in a reputable journal unless these questions were dealt with. But in 1863 (and for some time thereafter) no one dared to challenge the most eminent scientist of the day, and Kelvin's statements were accepted virtually as gospel truth. When Kelvin said that

> It is quite certain that a great mistake has been made—that British popular geology at present is in direct opposition to the principles of natural philosophy.

no one asked whether it was the principles or the assumptions he had made that created the "opposition."

In order to understand what happened after Kelvin's estimate was published, we need to report one more calculation that he made, a calculation for the age of the sun. The principle of universality tells us that energy must be conserved on the sun just as it is on the earth. The sun is clearly sending out energy—you can put out your hand and feel the heat. As he did for the earth, Kelvin asked the question, Where does that energy come from and how long can it last? We'll discuss some of the details of his reasoning in chapter 13, but the bottom line is that his models of the sun led him to the conclusion that it could have been shining for no more than 20 million years—about ⅕ of his estimate of the age of the earth.

So during the period after 1863, while geologists were scrambling to see if they could live with an age of the earth near 100 million years, Kelvin was concerning himself with the problem of reconciling his two estimates of the age of the solar system. He was, in other words, trying to find some way to modify his model so that the age of the earth would coincide (at least roughly) with the age of the sun. Modern physicists call this "pushing the parameters."

You can get some sense of what he was doing by thinking about our analogy of the ball rolling down the hill. Suppose that when we compared our predictions to the actual measured time of descent, we found that the prediction was five times too long—that we expected the ball to come down in fifty seconds and it actually took only ten. We would then start to modify our model to accommodate this datum. We might, for example, see what happened if we assumed that the friction on the slope was less than was originally thought. We would, in other words, "push the parameters" of the model to see if we could make the prediction coincide with the experiment. We couldn't do this arbitrarily, of course. The amount of "pushing" is limited by other experiments that could be done to determine the amount of friction on the slope. Nevertheless, there is usually a fair amount of leeway in the details and the numbers used in a theoretical model. There would normally be a range of frictional forces that would be considered reasonable and would fit with other information about the hill. There is nothing wrong with favoring those values that tend to push your predictions toward what you believe the right answer to be. But it is extremely dangerous to forget that that's what you're doing.

After the original publication of "On the Secular Cooling of the Earth," Kelvin kept using new data to push his estimate of the planet's age down to agree with his estimate of the age of the sun. By 1868, he abandoned this first tolerant attitude and was calling 100 million years the absolute maximum age the earth could have. In 1876, he lowered this upper limit to 50 million years, in 1881 he put it at 20 million to 50 million years, and by 1897 he picked 24 million years as the best estimate of the earth's age, with a possible spread from 20 million to 40 million. By the end of the century, then, he had reconciled his two estimates of the age of the solar system.

In so doing, of course, he was flying directly in the face of the geologists. By and large, once geologists started thinking about the problem of a limited lifetime of the earth, they saw the point of Kelvin's arguments and, indeed, by the end of the century they were prepared to accept his original age of 100 million years. Unfortunately, as we have seen, by that time Kelvin was proposing something quite different. The last part of the nineteenth century, in fact, gave us a sort of dog-and-pony show in which the geologists kept bringing their estimate of the earth's age down, and the physicists kept lowering their upper bounds on what the age could be and still be consistent with the "principles of natural philosophy."

Reading what the geologists were saying at that time, one

sees a growing frustration with the physicists. It wasn't that physicists looked at the geologist's evidence for an old earth and rejected it. They simply refused to *look* at geological evidence. For the physicists, most of whom had never been in the field or looked at a rock, there seemed little point in even considering anything so mundane as the observational evidence. The geologists simply *had* to be wrong, because their conclusions disagreed with Kelvin's model of the earth. Nor was the conflict helped when in 1869 a physicist such as P. G. Tait wrote in the *North British Review* in Kelvin's defense:

> Let us hear no more nonsense about the interference of mathematicians in matters with which they have no concern; rather let them be lauded for condescending from their proud preeminence to help out of a rut the too ponderous waggon of some scientific brother.

Against this sort of attitude we find geologists struggling to assert that they, too, possessed data that had to be taken into account by other scientists. Here is Sir Archibald Geikie, director general of the British Geological Survey.

> We must remember that the geological record constitutes a vo luminous body of evidence . . . which cannot be ignored, and must be explained in accordance with ascertained natural laws. If the conclusions derived from the most careful study of this record cannot be reconciled with those drawn from physical considerations, it is surely not to much to ask that the latter should also be revised.

As far as I can see, the first person to start talking sense in this impasse was a man named John Perry, a mathematician, engineer, and (perhaps most important) a former assistant of Kelvin's. In 1895, he published an article in *Nature* that, for the first time, raised the issue of the validity of Kelvin's model of the earth from a physicist's point of view. He showed that by dropping part of Kelvin's second assumption (p. 139) and assuming that the earth was not homogeneous throughout, it was possible to get estimates of the age of the earth as high as a billion years or so. He found that if he assumed that the material at the center of the earth conducted heat better than that at the surface, the kinds of calculations Kelvin had done yielded very different results from those Kelvin had published.

What is important here is not that Perry was right about the properties of materials at the center of the earth—he wasn't. What is important is that he showed that it was possible to argue that the earth is much older than Kelvin said without rejecting the "principles of natural philosophy." All that has to be rejected is some assumptions that had been made to simplify the calculations. In modern language, Perry showed that the Kelvin calculation was not "robust"—changing the assumptions a little changed the final result.

Radioactivity

It would have been interesting to see how the argument about the age of the earth would have played itself out had the battle been fought along the lines suggested by Perry. Kelvin had, indeed, used a grossly oversimplified model of the earth's interior. And while it is certainly true that the Laws of Thermodynamics describe the cooling of a planet, it is equally true that the presence of structure inside a planet makes the calculation of the time involved a great deal more complicated than appears at first.

But for better or worse the battle was not to be fought over Kelvin's second assumption, but over his first. The end of the nineteenth century saw the discovery of radioactivity. Marie Curie found new radioactive elements in ores from mines in Czechoslovakia, and it was realized that radioactive elements were widely distributed throughout the earth. And although the mysteries of radioactivity were still being solved during the first few decades of this century, by 1900 it was already becoming clear that this new phenomenon must play a major role in the discussion of the age of the earth.

The point is that radioactive materials are sources of heat. This heat isn't released all at once, but slowly, over thousands or even millions of years. Despite its slow release, the amount generated by radioactivity even in ordinary rocks is substantial. For example, if you could capture all the energy emitted by the radioactive elements normally found in a block of granite that you could hold in your hands, over a few million years it would be enough to melt the rock completely!

Most people think of radioactive elements as being of rare and exotic occurrence. In fact, they are quite common. Uranium is much more abundant in the earth than such familiar things as mercury and silver. The recent concern over the presence of radon in American homes is an illustration of this fact. The radon that seeps out of the ground and into homes is produced by the radioactive decay of uranium in underground rocks, and decay that produces radon also produces heat.

If the heat generated by radioactivity in all the rocks near the surface of the earth is added up, the sum is prodigious. The pioneers in the new science of nuclear physics realized that this heat constituted a source of energy within the earth, and that its existence negated Kelvin's postulate that all the heat flowing out of the earth was due to cooling from an initial hot state.

Of course, the young scientists were understandably leery of bearding the old lion. Ernest Rutherford, later a Nobel laureate and something of an old lion himself, gave the following account of his first confrontation with Kelvin.

> I came into the room, which was half dark, and presently spotted Lord Kelvin in the audience and realized that I was in for trouble at the last part of my speech dealing with the age of the earth. . . . To my relief, Kelvin fell asleep, but as I came to the important point, I saw the old bird sit up, open an eye, and cock a baleful glance at me! Then a sudden inspiration came, and I said that Lord Kelvin had limited the age of the earth *provided no new source of heat was discovered.* That prophetic utterance refers to what we are now considering tonight, radium! Behold! The old boy beamed upon me.

Beaming or no, Kelvin went to his grave maintaining that the nuclear physicists had not found a continuing source of heat within the earth, and he never published a retraction of his estimates of the planet's age.

The Resolution of the Conflict

The discovery of radioactivity changed the way people thought about the age of the earth. It not only swept away Kelvin's limits, it allowed geologists to go back to their basic task, which is the exploration of Deep Time. It also gave us (as we shall shortly see) a direct way of dating the earth and its rocks, with the resulting present estimate of 4.6 billion years.

Although Kelvin's calculations were discarded by scientists early in this century, this episode continues to fascinate members of the profession. Recently Frank Richter, a geophysicist at the University of Chicago, took a fresh look at the problem and showed that even had radioactivity not done the job of revision, the complex internal structure of the earth would have. Using the theoretical tools available to modern geophysicists, he showed that motions in the earth's interior would by themselves require a change in Kelvin's estimate of the age of the earth. Richter's conclusion: Even if you assume that all the heat coming out of the earth today results from cooling, and even if you neglect radioactivity completely, you can still "push the parameters" enough to predict an age that comes to billions of years. Thus, *both* of Kelvin's assumptions about the earth were wrong and contributed to his low estimate of the earth's age.

What does the Kelvin episode teach us? The obvious (and somewhat trivial) lesson is that it pays to look closely at the assumptions that go into any calculation. The bald statement that "science says X" (or its modern form, "the computer says X") must always be questioned. The story shows dramatically that an elegant (and correct) calculation based on faulty premises will give just as wrong an answer as a simple mistake in addition.

Still more important from our point of view, the great debate over the age of the earth that began with Hutton and ended with the discovery of radioactivity marks the period when the principle of universality began to be applied in time as well as in space. The arguments of both Kelvin and his opponents were based on the assumption that the same laws of nature that operate today operated at all times in the past. Their differences had to do with *which* laws they chose to consider, not with whether those laws apply only to a modern period. From this point on the principle of universality became an unconscious part of the folklore of science, so that even those who work at the very edges of the imaginable today rarely think about— much less question—whether it is valid.

TWELVE

The Nucleus

I hear it in the deep heart's core

—WILLIAM BUTLER YEATS
"The Lake Isle of Innisfree"

HERE ARE TWO senses in which the discovery of radio-activity eliminated the conflict between the principle of universality and the evidence of the earth's great age. First, as we have noted, it supplied our planet with an internal source of heat, thereby sidestepping Kelvin's arguments for a young earth. Second, as we shall see shortly, the phenomenon provides us with a clock which we can use to date the earth with a rather astonishing degree of precision.

The extension of the principle of universality throughout space depended on our ability to detect the "fingerprints" of atoms in the light they emit. This emission process, as we saw, involved electrons shifting around in their orbits in the atom. The extension of the principle of universality through time, on the other hand, depends on the presence of radioactivity, a phenomenon which is associated with the nucleus of the atom, not with the electrons. This split—electrons equal space, nucleus equals time—may seem a bit too easy, but it reflects an important fact about the way that matter is organized, as we shall see when we learn a bit more about the nucleus.

Just as we had to know about the atom to understand the work of Fraunhofer and Lockyer, we have to understand the nucleus to know about the people who came after Kelvin. Those who already have a firm grasp of radioactivity and the various forms of nuclear energy, or those who wish to follow the story and skip the details, can read the summary at the end of this chapter and move on.

The Nucleus

The first thing to grasp about the nucleus is how unimaginably small and dense it is. If the nucleus of an ordinary atom were a bowling ball sitting on a table in front of you, the electrons that constitute the rest of the atom would be like a handful of sand scattered over the rest of the county in which you live. The atom is almost all empty space, with the nucleus occupying a tiny dot at the center.

It is the electrons that give the atoms its size: They define its boundaries and its extension in space. The nucleus is all but lost in the emptiness at the heart of the atom. On the other hand, almost all of the mass of the atom is packed into the nucleus. It is the nucleus that gives the atom weight and heft. In a typical atom less than 0.1 percent of the mass resides in the electrons.

Several important consequences follow from this division of function between the electrons and the nucleus in an atom. For one thing, the scale of energy in the diffuse, spread out electron cloud is much smaller than that in the tight, massive nucleus. This means that it is easier to move electrons around between orbits than to make changes in the interior of the nucleus. From this it follows that the energy that comes out of the atoms when electrons move around will be much lower than those that accompany the disruption of the nucleus. In fact, we have seen that quantum jumps of electrons between orbits produce visible light. The analogous processes in the nucleus produce the much more energetic stuff we associate with radioactivity.

The second important fact about electrons and nuclei is that it is unlikely that anything happening among the former is going to have much of an effect on the latter. The nucleus goes its own way, indifferent to what the electrons are up to. The sorts of behavior we normally associate with atoms—the emission of light, for example, or the formation of chemical compounds—arises from the interactions of the electrons. They either make quantum leaps between orbits within a single atom or they interact with electrons from other atoms as chemical reactions go on. What they do, however, has virtually no effect on what the nucleus is doing.

In the light of the analogy of the bowling ball and the grains of sand, this state of affairs isn't hard to understand. The emission of light and the formation of chemical compounds involve the activities of grains of sand out near the county line. Obviously those activities will have little effect on what's going on in the bowling ball. This means that things like the location of an atom and the question of whether it is or is not incorporated into a mineral will have nothing to to with radioactivity, which depends only on the properties of the nucleus.

The nucleus of any atom is a mixture of two types of particles: protons and neutrons. These two particles are of roughly equal mass, and differ primarily in their electrical properties. The proton has a positive electrical charge, the neutron (as the name suggests) is neutral. In most nuclei, the number of these two particles is about equal, and if it is not, there is usually an excess of neutrons over protons rather than the other way around.

The protons and neutrons are arranged inside the nucleus in a miniature recapitulation of the Bohr atom. There are a series of allowed orbits called "shells." The protons and neutrons fill these shells just the way that electrons fill the Bohr orbits in the atoms. There are two separate sets of shells, one for protons and one for neutrons, with the proton shells slightly farther out than the corresponding shells for neutrons. This so-called shell model for the nucleus is sketched in figure 12.1.

The identification of the chemical species to which a nucleus belongs is made according to the number of protons present.

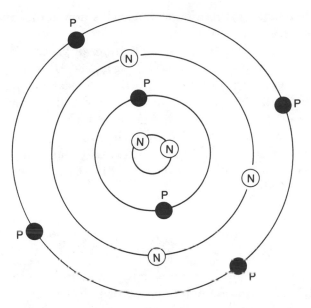

FIGURE 12.1

The rule is simple: For each proton in the nucleus there will normally be one electron in the Bohr orbits. It is these electrons — particularly those in the outermost orbits — that determine how an atom interacts with its neighbors. Since it is these interactions that we call the atom's chemical properties, we can say that it is the number of protons in a nucleus that determines the atom's chemical identity. For example, the carbon nucleus has six protons, the oxygen nucleus eight, the uranium nucleus ninety-two and so on.

For a given number of protons, nuclei can exist with an assortment of numbers of neutrons. Carbon, for example, has six protons. Normally, the nucleus will also contain six neutrons, for a total of twelve. We denote this type of carbon nucleus by the symbol ^{12}C. The *C* tells you that the nucleus is carbon, so that you know it has six protons, and the *12* tells you that there are twelve protons and neutrons, so simple subtraction tells you that the nucleus contains six neutrons.

A small percentage of carbon atoms in nature, however, do not have six neutrons. One type of carbon, for example, has a nucleus that contains six protons and eight neutrons. We call this carbon 14 (symbol: ^{14}C). As often happens when there is an excess of neutrons, this particular nucleus is not stable, and disintegrates over a period of time.

Atoms whose nuclei contain the same number of protons and different numbers of neutrons are called *isotopes*. All isotopes have the same chemical properties but different masses, corresponding to the different numbers of neutrons. All chemical elements have isotopes, and most have many. As with ^{14}C, most of these isotopes are unstable. In many cases, the isotopes are produced in the laboratory and are not normally found in nature. Still, most common elements possess at least two or three different stable isotopes.

Radioactivity

It would be wrong to think of the nucleus as a static collection of protons and neutrons resembling a bag of marbles. The nucleus is, on the contrary, a dynamic structure, always changing and shifting around. One reason for this lies in a simple fact: If all the protons have a positive electrical charge, they must exert a strong repulsive force tending to push them apart, disrupting the nucleus in the process. To overcome this tendency, there must be a force holding the nucleus together—a "nuclear glue." Physicists call this the "strong force," and think of it as generated by the exchange of hundreds of different kinds of elementary particles between the protons and neutrons in the nucleus.*

The crucial point is that *both* protons and neutrons participate in generating the strong force, whereas only protons suffer from the electrostatic repulsion. The neutrons can thus be said to play two complementary roles in the nucleus: they come between the protons, in effect diluting the electrical repulsion; and they contribute part of the glue that holds the nucleus together. This explains why heavy stable nuclei tend to have more neutrons than protons. But even with this help, the binding of the nucleus is always tentative and precarious. Sometimes, as in ^{14}C, even

* The modern view of forces as arising from the exchange of particles, and a historical survey of how that view arose, is given in my book *From Atoms to Quarks* (Scribners, 1980).

160

the extra neutrons aren't enough to overcome the disruptive forces that arise in the nuclear maelstrom, and the nucleus disintegrates.

When this disintegration occurs, the nucleus emits particles. We call these particles *radiation* and we say that the nucleus is *radioactive*. It was the discovery that there are processes in which nuclei emit very energetic particles that suddenly caught the attention of scientists around the turn of the last century and led to the resolution of Kelvin's puzzle about the earth's age. It was found that there were three different types of radioactivity — three distinct (but then mysterious) types of radiation that came from some nuclei. As if to emphasize the mysteriousness of the whole business, these were called (respectively) alpha, beta, and gamma radiation.

Today, we understand what these types of radiation are and the processes that produce them in the nucleus. They can be summarized as follows:

ALPHA RADIATION

Certain nuclei emit particles composed of two protons and two neutrons bound together. Such a particle is called an alpha particle. It is also the nucleus of the helium atom. When a nucleus emits an alpha particle, we say that it undergoes alpha decay.

When alpha decay occurs, the nucleus that is left behind (the "daughter" nucleus) has two protons and two neutrons fewer than it did originally. In alpha decay both the mass and the chemical identity of the nucleus change.

BETA RADIATION

A neutron that is not bound up in a nucleus will also disintegrate. On a time scale of the order of eight minutes, the neutron will be transformed into a proton, an electron, and a massless particle called the neutrino. This process is called beta decay, and the electron was once called the beta particle. This name was given, of course, before physicists realized what this particular radiation was.

For various technical reasons, neutrons cannot undergo beta decay in most nuclei. In some, however, this restriction does not apply and the nucleus itself undergoes beta decay. One of its neutrons changes into a proton and a very energetic electron (beta ray) is emitted. The daughter nucleus thus has one more proton and one fewer neutron than it originally had, but approximately the same mass. Thus, beta decay is characterized by a change in identity, but not mass, of the nucleus.

GAMMA RADIATION

Gamma rays are nothing more than high-energy photons. The term is somewhat fuzzy, encompassing both "hard" X rays and photons more energetic than X rays. Gamma rays are emitted from a nucleus by a process analogous to those that produce visible light in the Bohr atom. A proton or neutron makes a quantum jump to a lower shell, and the difference in energy between the two shells is carried away as a photon (gamma ray). In this process, there is no change in the number of protons and neutrons in the nucleus, and hence both the chemical identity and the mass remain the same.

The three forms of radioactivity are summarized in figure 12.2.

Radiation as a Source of Heat

Now that we understand that radioactivity is nothing more than the emission of rapidly moving particles from atomic nuclei, we can understand how it supplies the earth with a source of heat. For the sake of definiteness, consider the alpha decay of the nucleus of an atom deep within the planet. The alpha particle comes out of the nucleus at high speed — typically a few percent of the speed of light. It quickly leaves its home atom and enters the material of which that atom is a part. It starts colliding with atoms, bouncing through the material like a steel ball bearing through a pinball machine. In each collision, the alpha particle slows down a little. Part of its energy is transferred to the atom it hits, which moves faster after the collision than it did before.

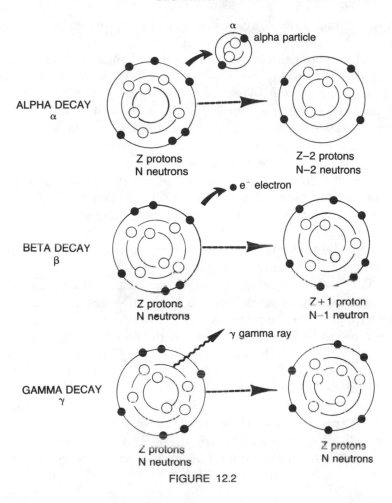

ALPHA DECAY
α

Z protons
N neutrons

Z−2 protons
N−2 neutrons

BETA DECAY
β

Z protons
N neutrons

Z+1 proton
N−1 neutron

GAMMA DECAY
γ

Z protons
N neutrons

Z protons
N neutrons

α alpha particle

e⁻ electron

γ gamma ray

FIGURE 12.2

We perceive a material in which atoms are moving quickly as being hotter than one in which they are moving slowly. The rattling around of the alpha particles makes itself known to us as an increase in temperature in the material in which it occurs. Radioactive materials are "hot" in both the literal and figurative sense. They give off radiation, which makes them "hot" in the slang of the nuclear technician, and they also attain high temperatures.

If there is an appreciable number of unstable nuclei in the material, this heating will go on for a long time. It will provide a continuing source of heat inside the earth, for example. Once we understand this, we can see that in a certain sense Kelvin was right about the earth. Radioactivity cannot provide an inexhaustible source of energy—once all the unstable nuclei have decayed the party is over and the earth must start to cool off. The question, then, is not *whether* the earth will eventually give up all of its heat to space, but *when* it will do so. As he pointed out, the lifetime of the earth cannot be infinite.

The length of time it takes for unstable nuclei to decay is measured in terms of the *half-life* of the particular isotope in question. Half-life is defined as the time it takes for half of the original complement of nuclei to decay. If there were one thousand nuclei in a sample right now and three hours later we find that five hundred had decayed, we would say that the half-life of that particular nucleus is three hours.

Half-lives for radioactive elements can range from microseconds to billions of years. There is no set of systematic rules that governs half-lives, nor do we yet know enough about nuclear structure to be able to predict what the half-life of any isotope should be. We can, however, measure any half-life with a high level of precision.

From the point of view of the age of the earth, the most interesting element is uranium, whose most common isotope is ^{238}U (92 protons, 146 neutrons). This isotope, which makes up the bulk of naturally occurring uranium and is actually quite common in the rocks forming the earth's crust, decays by alpha emission with a half-life of 4.5 billion years. This means that roughly half of the ^{238}U nuclei that were present when the earth was created have decayed, and that another quarter will decay in the next 4.5 billion years. While these decays are going on, the daughter nuclei of uranium, themselves unstable, also decay, giving rise to a chain of radioactive nuclei whose emission add to the heating effect. Eventually, of course, this source of heat will be exhausted. However, if you think the earth is only 20 million years old, a source of heat that is good for billions of years is for all practical purposes inexhaustible.

I am not aware of any reliable calculations of the time it will take the earth to cool, given its complement of radioactive elements, but the example of ^{238}U shows that this cooling time will probably be in excess of 10 billion years. In the next chapter we will see that this is comparable to the lifetime of the sun, so that the question is academic at best.

Radioactive Dating

The phenomenon of the radioactive half-life can also provide us with a means of measuring the age of objects, be they man-made or natural. There is no need to go into detail here about how the scheme works.* The basic notion is this: If for some reason you believe that you know how many nuclei of a particular isotope were present when an object formed, and if you know how many nuclei are present today, then you know how many half-lives have passed since that formation. If (as is generally the case) you know what the half-life of the isotope is, then you know the time of formation as well.

Take ^{14}C dating as an example. If you believe you know how much of this isotope was in the atmosphere in the past, then you can date artifacts made from organic materials like wood, cloth, or leather. The idea is that as long as the tree, plant, or animal that provided the material was alive, it was processing carbon from the atmosphere and incorporating that carbon into its own structure. When the tree, plant, or animal died the flow of carbon stopped and the complement of ^{14}C began to decay without being replaced by fresh carbon from the atmosphere. The half-life of this isotope is 5,730 years, so that after this time only half the original complement of ^{14}C would be present, after 11,460 years only a quarter, and so on. Thus, the death of the living thing can be fixed in time. From that we can form a pretty good estimate of the date when the artifact was made. This particular method of dating is widely used by archeologists to date the remains of old civilizations.

Geologists use a similar technique to date igneous rocks. The procedure is analogous to those given above, except that what is obtained is the date when the rocks last melted rather than

* A more complete discussion of radioactive dating and the various determinations of the age of the earth can be found in my book *Meditations at 10,000 Feet* (Scribners, 1986).

the date when a living thing died. Typical isotopes used by geologists, together with their half-lives are potassium-40 (1.3 billion years), rubidium-87 (47 billion years), uranium-235 (710 million years), and uranium-238 (4.5 billion years). All these isotopes are used to date the oldest rocks and to give the best current estimate of the age of the earth, which is 4.6 billion years.

Energy from the Nucleus: $E = mc^2$ and All That

When Albert Einstein published his special theory of relativity in 1904, I doubt whether he realized that he was laying the groundwork for a major extension of the principle of universality, but that was one of the outcomes of his work. The famous equation $E = mc^2$ tells us that matter is itself a form of energy —that there are processes in nature in which mass disappears and energy is created in its place. It is this type of energy that is responsible for radioactivity and thus, ultimately, for the heating of the earth.

People tend to view the equivalence of mass and energy as a kind of far-out product of the mind of theoretical physicists. They imagine that it has nothing to do with the real world. Nothing could be farther from the truth. An appreciable portion of the electricity in the United States is generated by nuclear reactors. These are devices in which matter is routinely destroyed and turned into energy that ultimately lights our homes and cooks our food.

The other aspect of the equation—the conversion of energy into mass—is also demonstrated daily, although not in a way that has such immediate consequences for most people. In giant particle accelerators located in laboratories all around the world, protons and electrons are routinely brought to speeds near that of light and then allowed to smash into targets. In the collision, some of the energy of motion in the moving particle is converted into mass, and new particles are formed. In other words, before the collision there is no matter present; after the collision there are particles just as real—just as massive—as the original proton. The mass that goes into these particles is made from the energy carried by the original proton.

This process is rather hard for many people to visualize because there seems to be something different, something special, about solid matter. But to the physicist, all matter, even our own "too, too solid flesh" is just one more form of energy, not fundamentally different from the stuff that flows through electrical wires and lights your lamps. Though it may seem strange, the physicist has decades of experimental verification to back up his point of view.

RADIOACTIVITY AS AN EXAMPLE OF MASS-ENERGY EQUIVALENCE

Suppose you performed the following experiment: First, you weigh a nucleus that is about to undergo alpha decay; then, you wait until the decay occurs and weigh both the daughter nucleus and the alpha particle. What you would find is that the sum of the weights of the daughter and the alpha is slightly less than the weight of the original nucleus. It is this slight difference that is transformed, according to Einstein's equation, into the energy of motion of the alpha particle. The heat that we associate with radioactivity is therefore derived from the destruction of a small amount of matter in each nuclear decay. And although we have talked here only about alpha decay, exactly the same holds true for beta and gamma decay as well.

To sum up, the ultimate source of energy for the earth's heat is the mass of radioactive nuclei. For each little bit of heat the earth gains, it loses a little mass. There is, indeed, no free lunch in this world!

FISSION

The power in nuclear reactors and in some nuclear weapons comes from a process known as nuclear fission. In this process, a nucleus is broken up into two or more fragments. If the mass of all the fragments is less than the mass of the original nucleus, fission must produce energy, usually in the form of energy of motion of the fragments. If the mass of the fragments is more than that of the original nucleus, that nucleus can still undergo fission, but energy has to be added to make the reaction proceed.

The nucleus that undergoes fission in commercial nuclear reactor is an isotope of uranium — ^{235}U. When a neutron hits this particular nucleus, the nucleus undergoes fission, producing energy and debris that includes several more neutrons. These neutrons can then go on to hit other nuclei, which produce more energy and still more neutrons, and so on. The result: a chain reaction that can cause a tremendous explosion (in a weapon) or, if controlled, used to generate electricity (in a reactor). In both cases, the ultimate source of energy is the destruction of a small amount of the original mass of the ^{235}U.

FUSION

In the larger universe, on the other hand, by far the most common source of nuclear energy is that generated by fusion. In the fusion process, two nuclei come together ("fuse") to form a third. If the mass of the final nucleus is less than the combined mass of the first two, then the difference in mass will appear as energy of motion, either of the final nucleus or of other particles that may be created during the reaction.

Consider one simple example of a fusion process. Suppose that two protons collide with each other. One possible outcome is illustrated in figure 12.3. A nucleus is formed along with an electron and a massless particle called the neutrino. The nucleus contains one proton and one neutron. This particular nucleus is an isotope of hydrogen known as deuterium, and the nucleus itself is sometimes called the deuteron, as if it were a single particle.

In this reaction, the masses of the final particles add up to less than the mass of the initial particles, so energy is created. The energy is manifested in the motion of the deuteron, the electron, and the neutrino. As we shall see in the next chapter, this reaction plays a crucial role in the generation of energy in the sun and other stars.

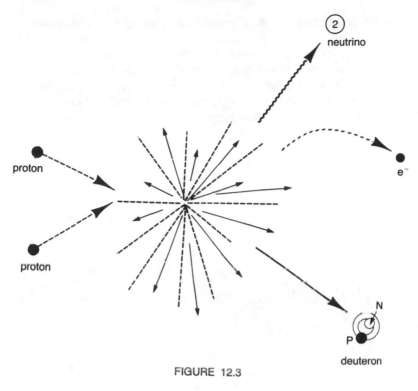

FIGURE 12.3

Summary

The nuclei of atoms are made up of protons and neutrons in roughly equal numbers. The number of protons in the nucleus determines the chemical identity of the atom; the combined number of protons and neutrons, its mass. Atoms with the same number of protons but different numbers of neutrons are said to be isotopes of each other.

There are three types of radiation given off by radioactive nuclei. They are called alpha, beta, and gamma radiation. Alpha radiation is particles made of two protons and two neutrons bound together. Beta radiation is composed of fast moving electrons, gamma radiation is energetic photons in the X ray region and beyond. The energy of motion of these particles, when transferred to atoms in the material containing the radioactive nucleus, manifests itself as heat.

Mass and energy are interchangeable, following Einstein's famous equation $E = mc^2$. The ultimate source of radioactive heat

169

is the destruction of small amounts of mass in each radioactive decay. Other forms of nuclear energy are fission, in which a nucleus splits into two or more fragments, and fusion, in which two nuclei come together to form a third. In both cases, the masses that are there before the reaction are greater than those that exist after. This mass difference is converted into energy of motion of any particles that happen to be created in the reaction.

THIRTEEN

The Age of Stars

The fault, dear Brutus, lies not in our stars,
But in our selves.

— WILLIAM SHAKESPEARE
Julius Caesar, Act I, Sc. ii

IT DIDN'T TAKE long for scientists to realize that the existence
of radioactivity meant that the earth was much older than
anyone had dreamed. It took a bit longer for them to realize
that the same phenomenon could be invoked to show that the
stars are older still.

If the discovery of the Laws of Thermodynamics created difficulties for people who thought about the age of the earth, they
were absolutely devastating for those who thought about the
age of the sun. You can see why by going outside on a sunny
day. If you close your eyes and turn your face to the sun, you
can feel the heat beating down on you. The First Law of Thermodynamics tells us that the heat you feel is a form of energy,
and that this energy comes from somewhere inside the sun.

This simple sunbather's experience can, with a few instru-
mental modifications, be used to estimate exactly how much

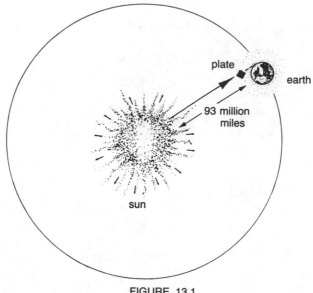

FIGURE 13.1

energy is generated by the sun. Imagine, as shown in figure 13.1, that we put a one-foot-square plate above the atmosphere of the earth. The plate is to act as an inanimate (and presumably more sensitive) surrogate for your face in the detection of radiation from the sun. We put it above the atmosphere because by doing so we can deal with energy from the sun directly, without having to worry about how it might be changed by passing through the atmosphere.

The amount of energy falling on the one-square-foot detector would correspond to a flow of a little more than one hundred watts—enough to power an ordinary light bulb. This may not seem like much, but if you add up all the energy falling on all the square feet on the surface of the earth, you obtain an energy flow of about 170 trillion kilowatts. By human standards, this flow of energy is prodigious—enough falls in half an hour to equal all energy generated by human effort in a year.

But, as illustrated in figure 13.1, the earth intercepts only a small fraction of the energy that the sun produces each second. Imagine a sphere centered on the sun with a radius of 93 million miles. The surface of the earth will then be part of that sphere

172

because the earth is also 93 million miles from the sun. Every square foot of the sphere will receive the same amount of energy from the sun, and the entire earth covers only a small fraction of the sphere. When you work out the numbers, you find that the earth intercepts only about four ten-billionths of the total energy streaming into space. From this fact, and from the known energy flux at the earth, it can be shown that the sun is radiating energy into space at the rate of about 10^{24} kilowatts—a number that is astronomical in every sense of the word.

One way to deal with this number is to suppose that the radiation arises because the sun started out hot and is just cooling off. Some simple mathematics shows that cooling simply could not produce this kind of energy outflow for any length of time. Something inside the sun must be replenishing its supply of energy. The question: Where does all that energy come from?

During the nineteenth century a number of ingenious calculations were done to provide an answer. One of my favorites was done by a French scientist who worked out the length of time a pile of anthracite coal as large as the sun could burn. Anthracite coal was the best fuel known at the time. Other investigators made use of other fuels—petroleum, for example, or a pure mixture of hydrogen and oxygen (the combination used to fuel modern rockets). In all these calculations, the lifetime of the sun came out much too short. With anthracite, for example, it was only barely longer than recorded human history. Obviously, the energy source of the sun had to be something other than a chemical reaction like burning.

Ruling out chemical reactions meant (at that time) that the energy somehow had to be found in the effects of gravity. For a while, astronomers seriously considered the possibility that there was a steady stream of meteorites falling into the sun. Each impact would produce heat, which was then supposed to be radiated away into space. This idea was taken seriously by a few people for a while (its proponents included Lord Kelvin), but it quickly encountered serious problems. The obvious objection was raised almost immediately—if all that stuff was falling into the sun, why wasn't it hitting the earth as well? Photographs taken during solar eclipses failed to show a flood of material falling into the sun, and it became clear that if there was enough matter in the form of meteorites to fuel the sun, its gravitational force would distort the orbits of the planets. So the meteorite theory of solar heating was quietly abandoned.

Then Lord Kelvin pointed out that it wasn't necessary to have material falling into the sun in order to have its heat generated by gravity. Heat could be obtained from gravity by having the sun contract.

That something can generate energy merely by contracting isn't obvious at first glance. You can convince yourself that it's true, however, by thinking of what happens when you drop something—a pencil, for example. When the pencil hits the floor, the energy of motion acquired during the fall is converted into heat. Each time you drop a pencil you add infinitesimally to the earth's store of heat and raise the earth's temperature by a tiny amount.

Now think about the idealized kind of contraction shown in figure 13.2, in which we imagine an object like the sun to be made of a core that doesn't change surrounded by a shell of material. If the outer shell falls down on the core, heat is generated. We have, in effect, dropped a large bundle of pencils, and the heat we generate in this process will be proportionately large.

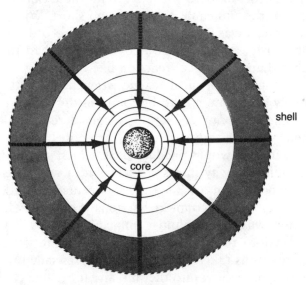

shell

core

FIGURE 13.2

Of course, the real sun does not consist of a core and an outer shell. Nevertheless, we can imagine that the overall contraction of the sun corresponds to a succession of shells collapsing, starting at the center and working its way to the outer periphery. When Kelvin actually worked out how much energy could be generated by this sort of shrinking collapse, he found that even the enormous outpouring of the sun could easily be supplied. He also found that the rate of collapse needed to produce this energy was far too small to be measured by techniques then available to astronomers.

Of course, gravitational collapse cannot go on forever. The sun must have a finite lifetime—like a wet dishrag, it can be squeezed only so much, and only so much energy can come out of a star during the slow process of collapse. Knowing the energy output of the sun today, it is relatively easy to estimate the rate of collapse, and from this to estimate the amount of time that a star like the sun can shine before it is "wrung out." For Kelvin, this lifetime was around 100 million years.

On the other side of the ledger, if we know how much energy the sun is putting out today, it is also easy to calculate how long it has been since the star formed from the collapse of a cloud of gas. In effect, one compares the energy of the gas cloud with the sun's present energy output and deduces how long this collapse has been going on. It was this sort of calculation that led Kelvin to the estimate of 20 million years for the age of the sun that we discussed in chapter 11.

The drawback of this calculation is that it is remarkably unforgiving. Unlike the calculation of the age of the earth, it is not possible to "push the parameters" and obtain an appreciably longer estimate. This explains why Kelvin kept trying to lower the age of the earth, and why he was so deaf to the arguments of the geologists.

Earth scientists, on the other hand, had clearly established that the planet was roughly 100 million years old, at least. For them, Kelvin's use of gravity to generate the sun's energy was little better than the old models that used anthracite coal. Neither allowed the earth to be as old as they knew it to be.

Such was the situation at the turn of the century. Confronted with a major contradiction that seemed to be unresolvable, scientists did what they always do in those circumstances. They ignored the problem and hoped that somebody would eventually find out how to make it go away. In the words of Sir Arthur Eddington (about whom more anon),

> About the beginning of the present century, the contraction theory was in the curious position of being generally accepted and generally ignored. Lord Kelvin's date of the creation was treated with no more respect than Archbishop Ussher's.

We owe our present knowledge of the workings of the sun to two prominent scientists. Arthur (later Sir Arthur) Eddington was a British astronomer who first realized that the new science of nuclear physics could supply a source of energy to take the place of Kelvin's gravity; and Hans Bethe, a German-American theoretical physicist who produced what is, in essence, our current understanding of how a star functions. Between them these men showed that the same laws that govern nuclear reactions in laboratories on earth, the same laws that supply the radioactive heating in the earth's interior, operate in the fiery centers of stars as well. With their work, the principle of universality was applied not just to the visible surface of the sun, but to its very core.

Unlike the debate over the age of the earth, the problem of the age of the sun seems to have generated very little rancor among working scientists. There was no question that the principle of universality was to be applied to this problem. Kelvin had, as we have seen, established the principle firmly during the nineteenth century. The problem was to find the processes that were actually going on in the sun and study them in the laboratory. I suppose it might have been possible to argue that the energy generation in the sun was so prodigious that no earthly process would work, and that this meant that new and unknown laws must govern the sun's interior. This would have been a violation of the principle of universality, but I can find no serious scientist who made that argument. Whether this was because everyone was witholding judgment while the new field of nuclear physics was explored or because everyone was overwhelmed by Kelvin's argument I do not know. What I do know is that until Eddington and Bethe had their say, most of their colleagues were content to adopt a wait-and-see attitude toward the solar energy problem.

Arthur Eddington was born in England in 1882 and reared in a strict Quaker home. Later in life, his deeply held religious beliefs caused him to take the unpopular path of becoming a conscientious objector when England went to war with Germany in 1914. He was a brilliant student, entering what is now the University of Manchester at sixteen and later winning entrance to Cambridge, where he was First Wrangler in 1904. This is the title given to the student who does best on the grueling Cambridge final exams, or "Tripos." After a stint at the Royal Observatory in Greenwich, he returned to Cambridge in 1913 and stayed there for the rest of his life. He died in 1944. He is perhaps best known for leading the eclipse expeditions in 1919 that confirmed Einstein's prediction that the path of light coming near the surface of the sun would be bent. This finding was the first major confirmation of the General Theory of Relativity, and marked the beginning of the public legend of Albert Einstein. It also established Eddington as a major figure in astronomy and lent credibility to his (largely successful) attempts to explain relativity to the general public.

But it is not Eddington's work in relativity that interests us here. It is the fact that he was the first person to put together something like our modern notion of what the inside of a star looks like, and to suggest that stars may be powered by some sort of nuclear energy. In order to make the case that nuclear energy must be what drives the sun, Eddington first had to demolish the "generally accepted but generally ignored" notion of gravitational collapse. He did this in a rather ingenious way, by thinking about an unusual star.

Astronomers had known for a long time that some stars in the sky do not shine constantly, but vary regularly in brightness over a period of weeks or months. The first of these stars to be studied intensively is called delta Cephei. In the nomenclature of astronomers, the name tells us that this star is the fourth brightest in the constellation of Cephus.* Stars are usually named by assigning letters from the Greek alphabet in order of bright-

* The constellation is named after an ancient king.

ness within a constellation, with alpha being the brightest, beta the next, gamma the third, delta the fourth, and so on. For example, alpha Centauri — the closest star to the earth — is the brightest star in the constellation of the Centaur.

Delta Cephei is a star that brightens and dims every five and a half days. This cycle is intrinsic to the structure of the star — indeed, we now know that it is a normal phase near the end of the life cycle of stars somewhat larger than the sun. There are many stars that brighten and dim like this, and they are known as "Cepheid variables" after delta Cephei.

In the late nineteenth century a series of careful measurements had established the fact that the length of time it takes a Cepheid variable to go through its bright-dim-bright cycle is related to its instrinsic brightness — the total amount of energy it is pouring into space. One use of this fact is to allow the Cepheid variables to play a role in determining distances of far away objects: We simply compare the total light emitted by the star (which we know from its pulsations) to the amount we actually receive here on earth.

But Eddington's use of this characteristic of Cepheid variables was different. He pointed out that since we know the energy output of delta Cephei today, we can do the same calculation for it that Kelvin did for the sun. This will tell us how much the radius of the star shrinks each year. The answer turns out to be about one part in forty thousand. He then proceeds:

> Now delta Cephei was first observed carefully in 1785, so that in the time it has been under observation the radius must have changed by one part in three hundred if the contraction hypothesis is right. Clearly changes of this magnitude could not occur without disturbing [the period of pulsation]. Does the period show any change? It is doubtful.

With this argument, Eddington showed that gravity, like the burning of anthracite coal, could not explain the fact that the sun (or any other star) continues to pour energy into space. With these two possibilities gone, what was left for him to try? Only the new and poorly understood energy of the atomic nucleus.

As we have seen, the process of radioactive decay can change one chemical element into another. At the same time, the differences in masses between the original and final nuclei appears as energy of motion of the particles given off. Eddington realized that the conversion of mass offered an almost limitless supply of energy to fuel a star, and suggested that this should replace gravity as the prime suspect in the mystery of the sun.

Actually, reading Eddington's papers with the advantage of 20/20 hindsight is a strange experience. We know the right answer, and we can watch him struggling with the problem until the answer seems almost within reach. Then suddenly, with the prize seemingly within his grasp, he turns up a blind alley.

Eddington realized that although all the nuclear reactions that he knew about were decays — that is, they produced small nuclei from larger ones — it was possible, at least in principle, that the opposite type of reaction should occur — that small nuclei should come together to form larger ones. Today, of course, we know this as nuclear fusion (see page 168). He also realized that a single helium atom (whose nucleus is two protons and two neutrons) was less massive than four hydrogen atoms, so that if there was a reaction that could convert the latter to the former, that reaction would release energy.

The problem was that no one had ever seen a fusion reaction in nature. Consequently, Eddington was forced to use indirect arguments.

> To my mind, the existence of helium is the best evidence we could desire of the possibility of the formation of helium. The [two protons and two neutrons] constituting its nucleus must have been assembled at some time and place; and why not the stars?

As we shall see shortly, it is precisely this sort of conversion of hydrogen to helium that powers the sun. But Eddington, with the answer almost in his hands, decided that this couldn't be it. Instead, he postulated that in some vague and undefined way, that matter in the sun is simply annihilated — turned completely into energy. He based this conclusion on the fact that older stars seem to lose some of their mass.* So near and yet so far!

---

* We now attribute this mass loss to the "solar" wind — a flow of particles out from the star into space.

After Eddington's work, scientists had been set in the right direction to solve the problem of the energy source of stars, even though the task had been left unfinished. Unfortunately, in his later years Eddington succumbed to a condition that seems to present particular risks for senior English scientists, and became embroiled in a kind of mysticism. He began multiplying and dividing the various constants of nature, seeing if certain numbers appeared more often than others. In this way, he hoped to discover deep truths about the universe. Best, I think, to draw a charitable veil over this attempt and remember his brilliant contributions to our understanding of the working of stars.

By the early 1930s, then, most scientists believed Eddington when he said that the sun was powered by the conversion of mass into energy, but no one really knew just how this mysterious conversion took place. Time went by, and gradually more and more of the mysteries of the nucleus were being unraveled. Nuclear physicists were getting better and better at their craft. But with their attention focused on the nucleus they weren't aware that the laws they were discovering might solve an old problem in astronomy. It was a classic case of the right hand not knowing what the left hand was doing.

Then in March 1938, a small conference convened under the auspices of the Carnegie Institution in Washington, D.C. As Hans Bethe described it later in his Nobel acceptance speech,

> At this conference the astrophysicists told us physicists what they knew about the internal constitution of the stars. This was quite a lot, and all of their results had been derived without knowledge of the specific source of energy. The only assumption they made was that most of the energy was produced near the center of the star.

The physicists, for their part, had been thinking about nuclear reactions and the energy they released. When they heard the astronomers talking about what was needed to explain the workings of the sun, they realized that they were very close to, if not actually at, the solution. The folktale that theoretical physicists tell is that Bethe actually worked out the theory of the nuclear reactions that go on in stars on the train from Washington back to Cornell. One can only wonder what would have happened if, instead of a leisurely train ride, he had taken a jet.

It was no fluke that Hans Bethe and no other finally cracked the problem of why stars shine. Born in Strasbourg in 1906, he was educated in Germany at a time when that country was the unquestioned leader of world science. He received his Ph.D. in Munich in 1928, but five years later the rise of the Nazis forced him to leave his native country. After a few years in England, he came to the United States and settled at Cornell University, one more great scholar bequeathed to us by a Europe sinking into darkness.

Throughout his career Bethe was a pioneer in the new science of nuclear physics, solving a number of difficult technical problems and quickly achieving a position of prominence. The picture of the way a star works that he finally developed can be understood by noting that at the center of a star, where temperatures are high and everything is moving fast, positively charged nuclei can overcome the normal electrical repulsion between them and actually come into contact. When this happens, a fusion reaction takes place, building the two smaller nuclei into a larger one.

The specific nuclear reaction Bethe proposed to power the stars is shown schematically in figure 13.3. Four protons interact sequentially to produce a helium nucleus and some other particles, with the mass deficit going to the energy of motion of the nucleus and other particles. These reactions involving protons are what supply the energy that ultimately leaves the surface of the sun in the form of light and heat. Since the proton is the nucleus of the hydrogen atom, we often say that the sun "burns hydrogen" to get its energy.

The important point about Bethe's work (and of many who have followed him) was that he took nuclear reactions that had been measured in the laboratory, extrapolated to the conditions that astronomers believed held in the core of stars, and showed that they could give the kind of energy output needed to run the stars. Can you think of a better example of the power of the principle of universality?

So the life story of a star, as told by the nuclear physicists, is

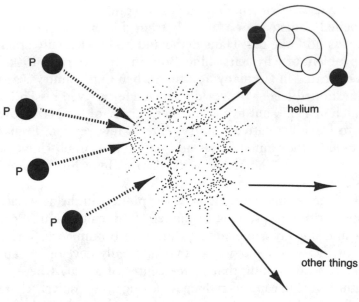

FIGURE 13.3

a simple one. As a cloud of interstellar dust starts to collapse under the influence of gravity, the central core starts to heat up. Eventually, this region heats up so much that protons initiate the reactions shown in figure 13.3, and the nuclear fires are ignited. The enormous outpouring of energy heats up the newly forming star and generates a pressure that, for a while, counteracts the inward tug of gravity. The star burns hydrogen, creating helium as an ash, and throws enormous amounts of energy into space.

There is plenty of hydrogen in the sun to serve as fuel, even though it is now being consumed at the rate of 600 million tons per second. The nuclear fires first ignited some 4.6 billion years ago, and will continue to burn for another 5 billion years or so. Other stars may burn up their hydrogen more quickly or more slowly, and will have correspondingly longer or shorter lifetimes. By and large, the sun can be thought of as a typical and quite undistinguished star.

When the hydrogen fuel gives out, the sun (and any other star) will die. In the case of the sun this death will be a slow, unspectacular evolution into a dimming star called a white dwarf. For larger stars, the end may be more spectacular — a supernova followed by a collapse into one of the imaginatively named end states associated with the stellar life cycle: black hole, pulsar, or neutron star. In all cases, however, we can be sure that the fuel will eventually give out and the star will stop shining.

So, as was the case in the debate over the age of the earth, when the dust had cleared it turned out that Lord Kelvin had been both right and wrong. It is incontrovertibly true that the law of energy conservation must apply to the stars as well as to the earth, and that the energy books have to be balanced. The fact that stars are pouring energy profligately into space today means that at some time in the future they will die.

On the other hand, it is also true that the sun and the stars will shine for a lot longer than Kelvin believed was possible. The lesson is obvious. We can understand our universe because the laws we discover here and now apply everywhere and for all time. But we had better be sure when we apply this principle that we know what those laws are, and that we know all of them.

FOURTEEN

The Expanding Universe: A Choice of Miracles

Everything that comes into being must arise either from what is or from what is not, and it is impossible [for it to come] from what is not. On this point all physicists agree.

—ARISTOTLE
Physics

The universe may be the ultimate free lunch.

—ALAN GUTH,
MIT

I LIKE TO THINK of the scientists who thought about the age of the earth and the stars as explorers in a new terrain— Lewis-and-Clarks of time, if you will. From their points of deepest penetration, they sent us back the message, "All's well—nothing new here." As far as they could tell, the familiar laws of nature governed everything back to the time of birth of the oldest stars.

But with these tidings new questions came to the fore. The laws of thermodynamics tell us that the stars will someday die, but they also tell us that the stars were born at some time in the past. The stars are not "from everlasting to everlasting," but from a well-defined (if distant) time in the past to another well-defined (but distant) time in the future. In extending the principle of universality into the past, then, we find that we must ask the question: What existed before the stars, and what laws of nature operated then?

There are only two ways that this sort of question can be answered. One can argue that the universe has always been pretty much as we now see it — that there was never a "before." In this case, either the laws of thermodynamics would have had to be different at some time in the past or there must be some source of energy in the past that we don't know about. Both of these options are, in a sense, antithetical in spirit to the principle of universality, for they imply that at some point in the past the rules of the game changed, and that we cannot, here and now, work out what the rules used to be.

On the other hand, if we accept the principle of universality, we are led to another, equally disturbing problem. If the conservation of energy tells us that the stars were born at a definite time in the past, then we have to ask where they came from. If we say that they condensed out of gas clouds, then we have to ask where the gas clouds came from. This process goes on until we come to the ultimate question: Where did the universe itself come from? How, in other words, can the laws of nature that operate in the universe here and now act to bring the universe itself into existence? Accept universality and you cannot avoid this question.

One of the most important debates that ever occupied science took place in the mid—twentieth century over the question of whether or not the universe had a beginning. The central observation that fueled the debate was the discovery that the universe is expanding.

The Expanding Universe

By the early 1920s the center of gravity of astronomy was shifting from Europe to the United States. Powerful new telescopes were being built here, and the crown jewel of the new instruments was the one hundred—inch telescope on Mount Wilson in California. Located in what was then a sparsely populated region west of Los Angeles, this telescope was the heart of the world's premier astronomical observatory. And to Mount Wilson one day in 1919, just when this new instrument was being put into operation, came a recently demobilized infantry officer by the name of Edwin Hubble, whose work was to revolutionize the way we think about the universe.

I have to admit that I have always had trouble figuring out how to react to Hubble. He was clearly an outstanding individual. An honor student at the University of Chicago, he lettered in track and basketball and was such a good amateur boxer that he had to make a serious career choice between the ring and graduate study. At one point, he was even touted as a Great White Hope — someone who could displace the black fighter Jack Johnson as heavyweight champion. Fortunately, Hubble decided to take a Rhodes scholarship and go to Oxford, where he studied law. After a short stint at lawyering in Kentucky (he didn't like it) he went back to Chicago to get his Ph.D. in astronomy. Almost as soon as he did so, the United States entered World War I, and Hubble enlisted as a private in the infantry. Rising to the rank of major by the end of the war, he was, at age thirty, finally ready to enter his chosen profession.

My mixed feelings about Hubble come from what I have been told and what I have read about his personality. Although he had a small circle of close friends who found him warm and congenial, to most of his colleagues he was cold, austere, and somewhat snobbish. For example, he was born in Missouri but spoke with an English accent throughout most of his adult life.* One writer summed up this aspect of Hubble's character by noting that "A photograph taken of him trout fishing, . . . staring into the camera, makes one feel for the trout." Perhaps the remarks in chapter 1 about the relation of humanity to genius apply in some measure to Hubble as well as to Isaac Newton.

In any case, coming to Mount Wilson just as the world's most powerful telescope was starting operation presented Hubble with an opportunity in which his drive and intellect could be used to their fullest extent. Because of the unparalleled resolving power of the instrument, Hubble was able to isolate individual

---

* Oxford sometimes does that to people. When I was a student there, members of the American community would say that someone who, like Hubble, adopted too many English mannerisms had "gone native."

stars in nearby galaxies. Working very carefully, he monitored the dimming and brightening in a number of Cepheid variables (see page 177). By comparing the light received at Mount Wilson with the amount of light emitted by the star, Hubble was able to show that the galaxies are millions of light-years from the earth—far outside of our own Milky Way. When he published this result in 1922, he resolved an old debate about whether or not other "island universes" exist. His answer: Not only do they exist, but most of the visible matter of the universe is in these collections of stars.

As he measured the distances to more and more galaxies during the 1920s, Hubble noticed a strange regularity. Almost all of the galaxies he was measuring were moving away from the earth, and the farther away they were, the faster they seemed to be moving.* The earth seemed to be located at the center of a vast array of receding galaxies. Out to the farthest distances he could see (about 6 million light-years), this relation seemed to hold. He summarized it in an equation

$$v = HD$$

which is to be interpreted as follows: v is the velocity with which a galaxy is receding from us, D is the distance to that galaxy, and H is a number now known as Hubble's constant. What the equation tells us is that if one galaxy is twice as far away as another it will be moving twice as fast, if it is three times as far away it will be moving three times as fast, and so on. This regularity, first discovered for a small number of nearby galaxies in 1929, is now known to hold out to the farthest reaches of the visible universe.

For over thirty years from the time that Hubble announced his discovery to the astronomical community, the puzzle of the motion of the galaxies was the subject of spirited debate. So long as the only thing we knew was that other galaxies were receding from us, there were two interpretations that fitted the facts. One interpretation, which came to be known as the Big Bang theory, was that the outward movement of the galaxies

---

*The details of the measurements Hubble made are given in many books, including my own *The Dark Side of the Universe* (Scribners, 1988).

was the result of a general expansion of the universe. The other, which went under the name of Steady State, postulated that as galaxies drew farther apart, new matter was created in the resulting voids, where new galaxies then formed. Advocates of both theories produced detailed mathematical models to justify their ideas, and both claimed (sometimes quite loudly) to be marching under the only true banner of universality.

Steady State versus the Big Bang

The Big Bang picture of the universe can be understood in terms of a simple analogy. Think of a lump of rising bread dough with a few raisins scattered throughout. Imagine further that you are standing on one of the raisins in the dough, looking at the other raisins. What would you see?

From your point of view, your raisin is stationary—you are not moving through the dough. On the other hand, if you looked at a neighboring raisin, you would see it moving away from you because the dough between you and it is expanding. In the same way, if you looked at a raisin that was twice as far away, it would be moving twice as fast as the nearer raisin, simply because there is twice as much dough to expand. No matter which raisin you looked at, it would be moving away from you, and the farther away it was, the faster it would move.

But this is just what Hubble saw when he looked out at the universe. The kind of velocity pattern summarized in Hubble's equation is a natural result of an overall expansion, and it is certainly reasonable to interpret the Hubble recession in this way. Of course, you have to remember that it is space (the bread dough) that is expanding—the galaxies are carried along in this expansion the way the raisins are carried along in the dough.

There is, however, a necessary consequence of this picture that many people found rather disturbing. If the universe is now expanding, you can think of "running the film backward" and watching the universe shrink as you go backward in time. The shrinking couldn't go on forever—at some point in the past everything would come down to a single point, with further shrinking impossible. In other words, if the universe is really expanding, *it must have had a beginning at some definite time in the past.* Explaining the creation of the universe out of . . . what? . . . presents a major conceptual problem that cannot be escaped if we adopt the Big Bang.

The first major attempt to work out the consequences of the Big Bang model was made in the post–World War II period by George Gamow, a Russian émigré theoretician, and Ralph Alpher, an American nuclear physicist. Gamow was one of the pioneers in the exploration of nuclear physics, and he had long speculated that it would be possible to use this branch of science to describe the evolution of the early stages of the Big Bang. His reasoning was simple: When the universe was very young, it was dense and hot—even hotter than the interior of stars. The collisions between particles were extremely violent, and positively charged nuclei could easily overcome the electrical repulsion between them and come together, initiating nuclear reactions of all kinds. Gamow felt that if we knew enough about these nuclear reactions, we could figure out how many atoms of each kind were made in the Big Bang. In his words,

> The relative abundances of various atomic species must represent the most ancient archeological document pertaining to the history of the universe.

The problem: how to read the document—how to translate our knowledge of the present abundances and scarcities of chemical elements into an understanding of the early stages of the universe.

An inveterate practical joker, Gamow added Hans Bethe's name to the paper that reported on these ideas even though Bethe had not worked on the paper (Alpher, Bethe, Gamow—get it?). The paper, appropriately enough, appeared in *The Physical Review* on April Fool's Day 1948. It marked the first attempt to put the Big Bang into rigorous mathematical form. Gamow imagined that the universe started as a hot, dense collection of neutrons which expanded, cooled, underwent nuclear reactions, and eventually became the familiar collections of stars and galaxies we see around us. He called the original collection of neutrons the "Ylem," from the Greek word for the chaos from which the world was born.

As it turned out, the basic assumption of this approach was wrong. It was implicit in Gamow's work that the entire complement of chemical elements was made during the first few minutes of the Big Bang, and that we have been living on our nuclear capital ever since. In fact, we now know that only light elements—up to helium (2 protons, 2 neutrons) and a little lithium (3 protons, 4 neutrons)—were made in the Big Bang, the rest being made later in the stars. Gamow and his coworkers discovered this basic flaw in their version of the Big Bang quickly, and the whole idea of tracing the evolution of the first, hot stages of the evolution of the universe passed into a kind of limbo for more than a decade.

But technical failures aside, we recognize that in 1948 physicists were not prepared to deal with the truly fundamental questions raised by the Big Bang. It's one thing to say that all the chemical elements come from the Ylem, but where did all those neutrons come from? On this question the scientists stood mute.

Meanwhile, in England, three young theorists were trying a totally different approach to the problem. They did not question the fact that galaxies are moving away from us, but they did question whether the Hubble recession necessarily implied a universal expansion from a definite beginning as the Big Bang's proponents seemed to think.

The three, all of whom went on to do important work on other topics, were Fred Hoyle, Thomas Gold, and Hermann Bondi. They first met during World War II, when they were working on the development of radar at a base near London. They apparently spent some of their spare time discussing cosmology. After the war, when they were back at Cambridge, the discussion became more serious. Recognizing that the Big Bang necessarily leads to the question of creation, Gold asked why the creation had to occur all at once, instead of being spread out over all time. The result of this line of inquiry was the Steady State model of the universe outlined above, in which continuous creation of matter produces new collections of stars in the voids left behind by the recession of already existing galaxies. At least one wag suggested that the model's primary purpose was to show that there would, indeed, always be an England.

Gold's argument has always reminded me of a sermon I heard as a child. The preacher asked us to imagine a situation in which he planted a seed in a pot, watered it, and had a plant grow to maturity in a couple of minutes. This would be a miracle. On the other hand, if the same thing happened over a period of several weeks or months, it would be just an ordinary event. His point, of course, was that even ordinary things are miraculous if you think about them in the right way. Seen in this light, the choice between the Big Bang and the Steady State universe represents a choice of miracles. Either the miracle happens all at once or it happens over a long period of time, but it's a miracle anyway you look at it.

Proof of the Pudding

We can build many universes in our mind. In fact, the task of the theoretical scientist is to construct possible universes—to tell us what *could* be. The task of the experimental and observational scientist is to see what *is*—to tell us which of these possible universes we actually inhabit.

The two possible universes—the Steady State and the Big Bang—are very different, and they make different predictions about the outcomes of certain types of observations and experiments. Hence they can be tested by the usual methods of science. These tests were done, and we will describe the outcome shortly.

What is striking about the early papers from this period is the degree to which the philosophical underpinnings of the theories were held to be important. Proponents of the Big Bang consciously adopted the position that the nuclear reactions that we measure here and now in our laboratory are the same as the reactions that took place when the universe was a few minutes old. Steady Staters, on the other hand, advanced what they called the "Perfect Cosmological Principle." Astronomers in the 1940s thought that the universe was uniform in space—that no matter which way you looked you would see the same thing. The Steady Staters argued that it should have a uniform appearance in time as well. Clearly, a "beginning" would violate this assumption, and therefore had to be rejected on philosophical grounds.

191

Both theories were cast in the language of universality, but it is clear that the Steady State violates the spirit, if not the letter, of the principle. Its proponents argued that the creation of mass really didn't violate energy conservation because the amount of matter being created at any given point is much too small to have been measured. You would, for example, have to watch a quart jar of water for a million years before even a single atom was added to it. The argument was that when we really understood the conservation of energy, we would find that it was violated by this infinitesimal amount, and that it is this amended version of the law that is universal, not the "ordinary" one we have now.

Despite this argument, it is clear that there is something anti-universal about the Steady State. There is something going on out there — galaxy building — that's not going on here. I look at the Steady State model as the most recent attempt to provide a worldview in which the heavens are different from the earth.

You should not infer from all this that philosophical reasoning is typical of modern astrophysics — it's not. This sort of argument arises only occasionally, when people are groping their way into a new area of inquiry with almost no data to guide them. Once the data start to come in, philosophical musings get shouldered aside and the scientific debate takes on a more familiar cast.

Throughout the '40s and '50s, the Steady State and the Big Bang were regarded as more or less equally matched combatants. At least they were presented side-by-side in textbooks and elementary lectures. Even then, however, the data seemed to lean in favor of the Big Bang. The reasons why are fairly simple.

If the Steady State model were correct, then you would expect to see young and old galaxies mixed more or less uniformly throughout space — as two older galaxies drew apart, a younger one would form between them. But it quickly became apparent that this uniformity is not found in nature. In fact, there do not appear to be young galaxies forming anywhere near the earth. This was cited by many authors as evidence against the Steady State hypothesis — pretty convincing evidence at that.

Unfortunately, there was a similar blemish on the Big Bang. If you think about the mental exercise of "running the film backward," you realize that a measurement of the current rate of expansion of the universe also supplies you with an estimate of the age of the universe. Hubble's original measurements indicated an age of the universe of about 2 billion years — younger than the age of the earth! Eventually, this was sorted out.* But for several decades the discrepancy between the age of the universe measured by the expansion and the age of the earth as measured by radioactive decay remained a predicament for the Big Bang.

These conflicts were just preliminary skirmishes in a war that ended abruptly in 1964. In that year Arno Penzias and Robert Wilson, working at Bell Laboratories in New Jersey, discovered that the universe is bathed in microwave radiation. For reasons that will become obvious in a moment, this was quickly recognized as incontrovertible evidence for the validity of the Big Bang explanation of the Hubble expansion. Since that time, the Steady State universe has had a shadowy kind of existence in astronomy textbooks as a foil for the Big Bang, but it has not been taken seriously by researchers.

You can understand the connection between the microwave radiation and the Big Bang by imagining a fire in a fireplace. After the flames stop, the coals are bright orange. As time goes by, the color fades to dull red and finally disappears. But even after the coals have stopped glowing, they are still giving off heat, as you can verify by holding your hand to them.

We interpret these observations by saying that hot objects give off radiation, and as a body cools the radiation it gives off shifts to longer wavelengths. The coals go from giving off yellow and red light (to make orange) to giving off only red to giving off invisible (but still detectable) infrared radiation as they cool. The kind of radiation given off at any time is a good indication of how long it has been since the fire was burning.

If the universe started off as a hot, dense material which cooled

---

* The solution to the problem posed by Hubble's original measurement is somewhat complex. It turns out that there are two different types of Cepheid variables, each with a different relation between period and brightness. By mixing the two in distant galaxies, Hubble got the wrong number for the distance, and therefore the wrong relation between distance (which he had wrong) and speed (which he had right). This problem wasn't straightened out until the early 1950s, when the universe once again became comfortably older than the earth.

off as it expanded, the radiation it emitted off would run through visible light to infrared all the way to microwaves. What Penzias and Wilson did was the electronic analogue of putting their hands to the "coals" of the universe to see what was coming off. The microwaves they found turned out to be exactly what you would expect to find in a universe that started expanding about 15 billion years ago. Since there is no natural way for such radiation to occur in the Steady State universe, this discovery led to the abandonment of that theory.

The Big Bang Today

With the demise of the Steady State universe, it was clear that the particular universe we live in is so arranged that we got our miracle all at once, in the beginning. This means that is is necessary to tackle the problem of the creation from nothing—or at least whatever it was that was there before the expansion started. But before this problem—the ultimate problem of cosmology—could be tackled, scientists had to work their way backward in time from the origin of the stars to the moment when it all started. This quest has occupied—perhaps even been an obsession for—many people who have worked at the frontiers of both physics and astronomy for the past two decades. Enough progress has been made for the vague outlines of the final answers to come into view.

The basic procedure, although it is not always simple to put into practice, is this. The first step is to work out what the temperature of the universe was at particular times in the past. This is a relatively straightforward job, involving little more than the calculation of the properties of an expanding gas. The next step is to decide what sorts of processes will be important at those temperatures: Do they involve atoms, nuclei, or something else? These processes are then studied in the laboratory (if possible) or in theoretical models, and the results of these studies are used to describe the properties of the early universe in detail. Finally, the results of this detailed description are used to predict properties of the present universe which are then compared with observations as a final test of the whole procedure.

This is indeed a textbook example of the way the scientific method is supposed to work. The crucial fact about it, from our point of view, is that it is based entirely on the principle of universality. The assumption that the laws of nature were the same when the universe was born as they are now is not even explicitly stated—it is simply assumed. The principle has been so successful in the past that scientists no longer even bother to acknowledge its use—it is simply taken for granted. We have indeed come a long way from that day when Isaac Newton walked in his orchard.

What emerges from the analysis of the early universe is a picture in which a number of relatively quiet and placid periods of expansion take place, interspersed with period of sudden change. The changes occur when certain critical temperatures are reached, and can be thought of as being analogous to the freezing of water at thirty-two degrees Fahrenheit.

The best way to understand the process of sudden change in the history of the universe is to consider the most recent episode. This involved the formation of atoms a few hundred thousand years after the beginning. At this point the temperature of the universe had become low enough so that if an electron fell into orbit around a nucleus, subsequent collisions would not be sufficiently violent to tear it off again. Before that time, no atom could survive. Even if one formed, everything was moving around so fast that the first encounter with another particle (a nucleus or an electron, for example) would break it apart. After a few hundred thousand years, once an atom formed it could stay together. The "punctuation mark" at a few hundred thousand years, then, was the formation of atoms out of a cloud of electrons and nuclei. Thereafter, the universe has continued to expand; galaxies, stars, and planets have formed; and life has developed in at least one place. But the fundamental unit of matter is still the atom, so one can say that the last 14,999,900,000 years have seen no fundamental alteration, no further punctuation mark.

We know how to put atoms together and take them apart—the process can be seen in a common device like the fluorescent lamp. Consequently, there is no mystery about the conditions that are necessary to create and destroy atoms: We can even recreate these conditions in our laboratories. The extrapolation to the early universe is thus not a difficult one. It involves little more than the assumption that atoms then behaved pretty much as they do now.

If the formation of atoms was the most recent punctuation mark in the history of the universe, the formation of the nuclei themselves is the next we could encounter if we went backward in time. About three minutes into the life of the universe, the temperature dropped to the point where nuclei could survive collisions. Before three minutes, if a proton and a neutron hooked together to form a rudimentary nucleus, a subsequent collision would tear them apart. After three minutes, the small nucleus would survive the collision intact and go on to add more particles to become a larger, more complex nucleus. Hence, in the period before three minutes matter was in the form of the elementary particles (like protons and neutrons), but after three minutes there are nuclei present waiting for the temperature to drop so that the electrons can hook on to form atoms.

Incidentally, this burst of nucleus building is short-lived for the simple reason that the Hubble expansion thins things out quickly. As a result, collisions that could lead to heavy nuclei being formed were rare. That is why only light nuclei form during the Big Bang, and why Gamow's first attempt to describe the early universe failed.

As was the case with the formation of atoms, we can reproduce in our laboratories the reactions that took place when nuclei formed at the three minutes mark. This period of "nucleosynthesis," as it is called, is another step in our story that we can reasonably believe.

Moving backward in time from three minutes, we see a sea of elementary particles getting progressively hotter. Collisions become more and more violent until, about ten microseconds from the beginning (ten *microseconds!*), the particles come apart into quarks.* We believe that quarks are the ultimate constit-

---

* The story of quarks and how we came to know about them is given in my book *From Atoms to Quarks* (Scribners, 1979).

uents of the particles that make up the nucleus. It has to be added that although we can indeed reproduce in our laboratories temperatures such as those that existed a few microseconds into the Big Bang, our current thinking is that we will never be able to see quarks as independent entities. In essence, the freezing of quarks into particles early in the history of the universe is a kind of one-way street; we cannot now unfreeze them. This means that our confidence in our knowledge of this stage in the Big Bang will have to rest (perhaps forever) on indirect, though copious, evidence.

You may be surprised to learn that even though we have traced the universe back to a few microseconds of the moment of creation, we still have some distance to go. Those first few microseconds contain not one, but three separate punctuation marks. They can only be understood in terms of the the new unified field theories, only recently developed by theoretical physicists. We look at these theories and their implications in the next chapter. What we shall see is that we stand today on the verge of the ultimate test of the principle of universality. Can we extend the knowledge gleaned from our laboratories not just to the beginning of time, but to whatever came before? Can we understand how the universe came into being from nothing? As the remark quoted at the beginning of this chapter indicates, some physicists think so. Perhaps you are interested in finding out why.

FIFTEEN

Creation:
The Ultimate Test

*We trace back events and come to barriers which close
our vistas. . . . They stand like gates of ivory and of horn;
portals from which only dreams proceed.*

—JOHN JOLY,
Irish geochemist, 1905

ONLY A FEW MICROSECONDS separate the decomposition
of matter into its most fundamental constitutents and
the event we have been calling the "beginning." The
closing of this seemingly infinitesimal gap is now occupying
the best theoretical minds in the world, and, as often happens,
the obstacles that seem to be the smallest are proving to be the
hardest to overcome.

If I had to pick out a single overall characteristic of the evolution of the universe, it would be the development of complexity from simplicity. The universe seems to get simpler as we
move backward in time. An atom, for example, is a relatively
complex structure—a conglomeration of protons, neutrons, and
electrons. A quark, on the other hand, is a simple structure—
indivisible and unstructured. Moving backward from one hundred

198

thousand years to ten microseconds, therefore, corresponds to finding matter in progressively simpler states until at last we find it in the simplest state we can imagine.

The universe continues to become simpler as we trace back through those first ten microseconds to the moment of creation. The only difference is that the changes involve the forces that mediate the interactions of matter with itself, rather than the form in which matter exists. Physicists know that there are four fundamental forces in nature. One of them, gravity, is familiar to us from our earliest experience. Electricity and magnetism, which are grouped together as a single kind of force, are also familiar from everyday experience. Two other forces, less familiar, operate inside the nucleus. The strong force acts to hold the nucleus together, while the weak force operates to produce certain kinds of radioactive decay. Anything that happens in the world happens because one or more of these forces is acting.

These forces appear to be very different. For one thing, they are effective over very different distances. Gravity acts across the depths of space from one galaxy to the next. The strong force, on the other hand, acts only across a single nucleus. The forces also vary greatly in strength. You can, for example, take a magnet that will fit into your hand and lift a nail, even though the entire earth is on the other side of that nail pulling downward with the force of gravity. This means that the magnetic force must be much stronger than gravity. Similarly, protons stay together inside a nucleus, even though the electrical forces between them tend to push them apart. This means that the strong force must be, as the name implies, stronger than the electrical force.

At the ten microsecond mark, when matter had been broken down into its simplest constituents, there were four forces acting, just as there are today. To a physicist, this fact represents a kind of complexity every bit as real as the complexity of the atom. The reason for this point of view is this: We need only one kind of force to make things happen in the universe. Any more are, in the physicist's eye, an unnecessary elaboration. Consequently, if we really believe that the universe should become simpler as we move backward in time, then we have to argue that there must be some way of reducing the number of forces.

The way that this is done is to find a way of looking at the universe in which two apparently different forces are seen to be nothing more than different aspects of a single force. When we do this, we say that the forces are to be regarded as "unified," and the theory that describes the single force whose two aspects we see is called a "unified field theory." In a sense, Newton's theory of gravitation was a unified field theory because it showed that two seemingly dissimilar forces (earthly and heavenly gravity) were actually the same force.*

Unification may seem an abstract notion, and it is true that the theories that I'm talking about are subtle as well as complex. But a simple analogy may simplify things for you. In winter, when you see ice and snow outside, you would be justified in supposing that ice is different from water. You could not see in the two substances anything to suggest that they were in any way related. But if you wait a while, you can see the ice melt. After that, it is obvious that there is a deep connection between the water and the ice—a connection that was not evident at first because of the low temperature.

In exactly the same way, theorists tell us that the fundamental forces look different to us because we live in a universe that has been expanding and cooling for 15 billion years. If we go back far enough in time, the argument goes, we would come to a temperature where the differences would disappear. Just as the ice melts so that it's basic similarity to water becomes manifest, so do the forces "melt" when the temperature is high enough. This melting process is what we have called unification. The only question to ask, then, is at what times the universe was hot enough for unification processes to occur.

It turns out to be fairly easy to calculate what these times should be. A general scheme of the first ten microseconds in the life of the universe is shown in figure 15.1. The first forces to unify during our journey back through time are the electro-

* A more complete description of the modern idea of force and the process of unification is given in my book *The Moment of Creation* (Scribners, 1983).

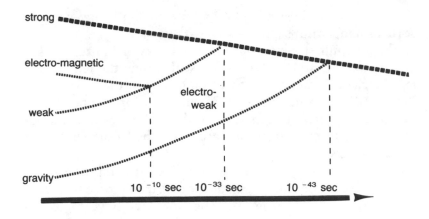

FIGURE 15.1

magnetic and the weak forces. At 10^{-10} seconds these unify into what is called the electroweak force.* Before that time, there were only three forces operating in the universe: the strong, the electroweak, and the gravitational. Afterward, there were the full complement of four. Thus, at 10^{-10} seconds the universe did indeed go from from a simpler to a more complex state.

And while it may seem incredible, we can actually create in our laboratories the conditions that existed in the universe at that time. At the European Center for Nuclear Research in Geneva, Switzerland, there is a large machine that accelerates protons almost to the speed of light, then makes the speeding protons collide head on. In these collisions so much energy is available that for a fleeting instant the conditions of the universe at that early time are established in a volume of space smaller than the proton. In this microscopic maelstrom, particles are created that have not existed since the beginning of time and evidence for their existence is registered in complex electronic machinery. The verdict of measurement is clear: The unification

---

* Throughout this chapter I will be using the so-called scientific notation to save space. The symbol 10^{-10} should be interpreted to mean "move the decimal place ten places to the left," so that we are talking about the forces unifying at 0.0000000001, or one-tenth of a billionth of a second.

theories do indeed describe the way that matter behaves at these sorts of temperatures.

The next milestone on our journey backward in time is supposed to occur at 10^{-33} seconds. At this time the temperature is so high that the strong force unified with the electroweak. Before 10^{-33} seconds there were only two forces operating in the universe, afterward there were three. The theories that describe this particular transition are called the Grand Unification Theories, or GUTs.

We cannot now, or are we ever likely to, reproduce in a laboratory the conditions that allow the strong force to become unified with the electroweak. Therefore, we have to use the GUTs is a somewhat different way. We begin by working out some of the consequences of a particular version of the theory for processes that take place a low energies. We check these predictions experimentally, verifying that the theory does indeed describe these processes and using the results of these experiments to pinpoint the various details of the theory. This tested and adjusted theory is then used to describe the early universe, and the consequences of this application are then compared with what we see in the present universe. It is only in this way that we can make comparisons between the GUTs and observation.

When this process is applied to the GUTs, the results are mixed. In the late 1970s, there was a great sense of triumph when a number of researchers found that the theories predicted that any antimatter that might have been around at the beginning would have been eliminated when the strong force split away, and that the universe must be composed almost entirely of ordinary matter. Since this is the kind of universe we see when we look out into the sky, and since no one had been able to suggest a good reason why things should be this way, the predictions of the GUTs were welcomed by cosmologists. This was, in fact, the first calculation that linked the early universe to new theories being spun out by elementary particle theorists, and can fairly be said to have initiated the great boom in cosmology of the 1980s.

On other fronts, the GUTs were not so successful, for they made a striking prediction that was never borne out by experiment. According to the GUTs, the proton, which we normally think of as being stable, should undergo a kind of radioactive decay. The proton lifetimes that the theories predicted were very long—much longer than the lifetime of the universe. Nevertheless, during the early 1980s a number of very sophisticated experiments were carried out in which the protons should have been seen to come apart if the simplest predictions of the GUTs were true. All these experiments came up dry—no proton decays were seen. Of course, the theorists could "push the parameters" (see chapter 11) to make their predictions of the rate of proton decay low enough so that the experiments couldn't have seen them. This "saved" the theories, but it doesn't fill one with a sense of confidence.

So the frontier of theories that are being tested by experiment, even indirectly, is at 10^{-33} seconds. The final milestone, at 10^{-43} seconds, is supposed to be where the gravitational force unifies with the strong-electroweak. Physicists call this the "Planck time," after Max Planck, the German physicist who was one of the founders of modern physics. Before that earliest time there was only one force operating in the universe—the ultimate in simplicity. It is what happened in that era that is of most interest when we try to understand how the universe came into being. How well are we doing in understanding this final frontier?

The theories that are supposed to describe the ultimate unification of the forces are sometimes referred to, with more than a little irony, as the TOE—Theories of Everything. They go by names like "supersymmetry" or "superstring." The "super" prefix simply means that the theory treats all four forces in a unified way. The last part of the name refers to some attribute of the particular theory being described. The superstring theory, for example, is one in which all particles (including quarks) are pictured as having a tiny stringlike structure buried inside them, with different particles corresponding to different types of vibrations of the strings.

All of these "super" theories share some common features. For one thing, they are extremely difficult to deal with in a technical, mathematical sense. Much more disturbing is that none of them—not one—has been developed to the point where any experimental or observational test of its basic premises is possible. I find this disturbing because these theories have been extant for almost a decade, and an entire generation of theoretical physicists have completed their training by working on them. These theorists have little sense of the complex interplay between calculation and experiment which is the central theme of our science. Some of my colleagues (perhaps reflecting our entry into middle age) shake their heads and say, "They're not physicists, they're just mathematicians!"

In any case, this is the state of knowledge of the early universe today. We can be fairly certain of the story in the period from 10^{-33} to 10^{-10} seconds; there are testable theories with a record of some success that take us from 10^{-33} seconds back to the Planck time; and the various "super" theories, which remain untested and perhaps untestable, can take us into whatever lies beyond. It does seem as if we are in a "so near but yet so far" situation, a minute fraction of a second away from the origin of the universe. Actually, this may not be the case. It may be that because of a rather unusual recent discovery, we can be assured that we now know all (or almost all) that we can know about the early universe, and that we can plunge right ahead into speculations about the creation without nagging doubts about whether or not new advances in theory will make our speculations obsolete.

Inflation and the Universal Loss of Memory

I walked into Alan Guth's office at MIT on a gray February afternoon. It was one of those days when the weather can't seem to make up its mind about whether it is snowing or raining or sleeting—a thoroughly dreary sort of day. Guth was talking to a student, but his youthful appearance, accentuated by his '60s style pageboy haircut and his standard uniform of tan slacks and tieless button-down shirt, would have made it difficult for a newcomer to tell which of the two was the professor. In fact, one of the striking things about Guth is that he just doesn't fit the popular image of a cosmologist. He could be a casually dressed stockbroker or a junior executive on the way up, but the bearded, scholarly look we expect to find in someone who spends so much time contemplating the origins of the universe just isn't there.

Appearances are deceiving. The truth is that in the early 1980s Alan Guth worked his way to a profound insight into what happened to the universe when it was 10^{-33} seconds old, an insight that both solved a number of old puzzles in cosmology and provided us with a new roadmap to guide our steps toward the creation. As shaped, refined, and molded by Guth and a number of coworkers, this insight is now embodied in what cosmologists call the "inflationary universe."

Alan Guth grew up in Highland Park, New Jersey. From an early age, he wanted to be a scientist or an engineer of some sort, and the only real choice he faced was deciding what kind. Like many scientists, he was strongly influenced by a teacher, in his case a man named Robert Landrum, who taught physics in his high school. It was no surprise, therefore, when he went off to MIT and emerged a number of years later with a Ph.D. in theoretical physics.

In the early stages in his career, he worked on problems in particle physics, following a thoroughly conventional career track. The early 1970s were a strange time in this field. The unified field theories had been discovered, but were being largely ignored. People were trying out a whole smorgasbord of research schemes, trying to break out of what was widely perceived to be a state of stagnation. Quantum field theory, the subject of Guth's research, was typical of this sort of work. Dealing with mathematical techniques that were supposed to describe the behavior of elementary particles, this field was solid, safe, and, to tell the truth, a little dull. During stints as a junior researcher at Princeton, Columbia, and Cornell, Guth tended his field theory vineyards, slowly building his reputation and advancing in a solid, respectable sort of way.

Then, in the fall of 1978, he started the transition from being a conventional field theorist to being a pioneer in the study of the early universe. In his mind, there were two events that marked the beginning of this passage. One was a seminar by Princeton physicist Robert Dicke, who discussed some of the problems then confronting cosmology. "I didn't do anything at the time,"

Guth says, "but I stored it all away." He began talking to colleagues about the subject, but couldn't quite bring himself to make the switch. "I resisted all the way," he says. "I was used to well-defined problems, just the opposite of cosmology. I felt I was moving onto shaky ground."

Then, the next spring, the final push came. Steven Weinberg, perhaps the most prominent theoretical physicist of our generation, gave a talk in which he described some of the work he was doing on the formation of the elementary particles during the first second of the life of the universe. Guth's reaction: "This makes the whole field respectable."

It was an interesting reaction, for it illustrates a very important point about the way the scientific enterprise is organized. At any time, various fields have a different status within the scientific community. Some are hot and "sexy," some are solid but staid, and some are beyond the pale. This categorization is to some extent a reflection of fashion, but mainly a collective judgement about what sort of research is likely to lead to new and exciting results. In the late 1970s the study of the early universe was making a transition from being beyond the pale to being a hot topic. Before that time it was felt (rightly, I think) that we just didn't have the tools to do anything useful, so that work in this area was likely to be purely speculative.

With the advent of the unified field theories this situation changed. Suddenly, the tools for investigating the early universe were available, and the cost-benefit analysis of working in the field went through an abrupt change. Now there were new territories to be explored, new paths to follow. The fact that someone like Steven Weinberg took the initiative was, in effect, a symbolic legitimation of the whole enterprise. If a senior theoretician thought it was worth doing, why shouldn't a younger man take the plunge?

So with Henry Tye, a colleague at Cornell, Guth started to think about applying the techniques of the Grand Unified Theories to the problems of cosmology. The pair's entry into this new field was not particularly auspicious. The first calculation they did was never published — they were "scooped" by another physicist. "We wanted to publish *something*," Guth said, "so we started to think about another aspect of the problem that nobody had worked on." About this time, he moved from Cornell to Stanford, continuing the collaboration long-distance.

In the process of trying to salvage something to publish, Alan Guth discovered the basics of the inflationary universe. The folklore of physics has it that the name comes from the severe inflation that was savaging the American economy in 1979, but Guth says that he can't remember the origin of the name.* As amended and improved by a number of other theorists, this is now the accepted version of the way the universe came to be the way it is.

Stripped of its mathematical complexity, the inflationary universe is not difficult to visualize, at least in principle. When the strong force separated from the others at 10^{-33} seconds, the universe was going through a "freezing" similar to the sort we see when water freezes into ice. When water freezes, we know that it expands — this is what causes pipes to burst during cold snaps. In an analogous way, when the strong force "froze" out from the others, the universe underwent a very rapid expansion.

But while the idea of an expansion coupled with a freezing may not seem unusual or exotic, the *amount* of expansion in the inflationary universe is. During this period the universe increased in size by no less than a factor of 10^{50} — that's a one followed by forty-nine zeroes! For reference, if your height increased by that much, you would stretch from one end of the observable universe to the other with your head and feet sticking out of the ends. At 10^{-33} seconds, the universe went from being smaller than the smallest elementary particle to being roughly the size of a grapefruit. This almost unimaginable expansion is what Guth called "inflation." And although I have presented the theory in a very sketchy way here, take my word for it that in its full-blown form it is a complex and technically difficult piece of mathematics.

It's hard to overestimate the impact that the concept of inflation has had on the world of cosmology. "The late '70s and

---

* I told him to make up a good story for future historians, but he's probably too honest to take my advice.

early '80s were a time of euphoria. Everything seemed to be working—falling in place," says Guth. Indeed, a number of old questions were answered then: Why is the universe made from matter instead of antimatter? Why is the microwave background radiation so uniform?* But from our point of view the most important thing about inflation is that it creates a boundary beyond which we cannot map the evolution of the universe in the kind of detailed way we can map what has happened since it occurred.

To see why this is so, it will be useful to think of an analogy. Imagine that you have one hundred rubber bands on a table, each twisted up and distorted in a different way. Now suppose you take each rubber band and stretch it out to the fullest extent possible. Once stretched (inflated), the rubber bands will all be essentially identical. You will be confronted with one hundred identical rubber bands and there is no way you could take one band and say which of the original one hundred twisted rubber bands it was. By stretching the rubber bands, you cause them to "forget" the state from which they originally came. In just the same way, inflation caused the universe to forget its initial state as well.

The inflationary universe creates for cosmologists a classic "good news—bad news" situation. The good news is that no matter how the universe started out—no matter how "twisted" it was at the beginning—it must wind up in its present state after being "stretched." Thus, the question Why is the universe the way it is? is simple to answer. The universe is the way it is because it underwent a period of inflation when the strong force froze away, and no matter how it started out the "stretching" associated with this inflation produced the universe we live in. In fact, you could probably go so far as to say that inflation implies that the universe we live in is the *only* universe consistent with the laws of nature, since these laws lead inevitably to inflation.

The bad news is that the inflationary epoch forms a barrier

* The way that inflation and the GUTs solve these problems, and why the problems are important, is discussed in detail in my book *The Moment of Creation* (Scribners, 1983).

for our march backward in time. If *any* initial conditions lead to the present universe, then there is no way that measurements made today can pick out the exact conditions that obtained before inflation took place. The problem is exactly analogous to trying to decide what a given rubber band looked like before the stretching occurred. We may be able to use the principle of universality to talk about the laws that governed the universe before inflation, but we can never find out much detail about that period.

To drive this point home, imagine that you wanted to compare two models of the preinflation universe. What you would do is to take the two, calculate what they predict for something we can measure today, then compare the predictions with an actual measurement to decide which model is most like the real universe. But because of inflation, both models will give exactly the same prediction for anything we can measure today. How, then, can we choose between them? The answer is: we can't.

That inflation creates a barrier to our further exploration doesn't prevent us from developing theories of how things began. It does, however, prevent us from choosing between competing theories on the basis of current measurements.

How Did It All Begin?

Given the likelihood that the correct theory of the universe will include the phenomenon of inflation, does it follow that we have to give up any hope of finding out how the universe was created? Provided that we are willing to settle for a general idea, rather than a detailed calculation that can be tested against observation, the answer is no. In fact, a number of audacious theorists have already entered the ring with theories of creation.

Go back to our analogy between the expanding universe and rising bread dough. We talked about "running the film backward" to estimate the time since the beginning of the expansion of the dough, and we spoke somewhat loosely about "creation" corresponding to the time when the it had contracted to a single geometrical point. We know, of course, that this could not happen. Running the film backward, we would indeed see the dough contract as we went back, but at some point the contraction would stop and a totally new process would start. We would see the bread start to come apart into its constituent parts: flour, eggs, yeast, and so on. In exactly the same way, those who worry about the origin of the universe assume that the current expansion has not gone on forever — that some time before the advent of inflation the universe was quite different from what it is today. The laws of nature were the same (or at least we assume they were), but the basic organization of things was different. The problem is to explain how the laws of nature took the universe as it was then and made it into the universe as it is now.

In most people's minds, the basic problem of creation is simply stated as, How do you get something from nothing? What, in other words, corresponds to the putting together of the flour, eggs, and yeast in our bread dough analogy? To understand how modern physicists deal with this question, you have to know what is in their minds when they use the term "nothing."

In modern physics, the vacuum is not simply the absence of matter. While it is true that in a vacuum there is no matter at any point *on the average*, the vacuum is actually full of activity. According to the laws of quantum mechanics, it is possible for particles and their opposite numbers, the antiparticles, to appear briefly out of nothing, then interact quickly so as to wipe each other out and disappear. This now-you-see-it-now-you-don't sort of event is called a "quantum fluctuation." The uncertainty principle tells us that so long as the particles appear and disappear quickly enough, we can never tell if the law of conservation of energy has been violated. In effect, the vacuum is full of elementary particle versions of Cinderella's coach — provided they get home before midnight, we'll never know whether they turn into pumpkins or not. Thus, "nothing," to a physicist, is actually a very complex and complicated system, not at all like the old notion of an inert emptiness.

The most likely forerunner of the universe is, of course, just this sort of complex "nothing." But how could something permanent arise from these epheremeal fluctuations? The general approach to answering this question is to (1) see if the momentary appearance of a small amount of mass could drain energy from another source (the gravitational field is the most commonly suggested), and (2) show that there is a process by which a fluctuation, once started, will continue to drain energy from that source and convert it into mass. This mass will then cause more energy to be drained, which will, in turn, create more mass, and so on. In these scenarios, the vacuum is actually unstable, so that a small disturbance leads quickly to large effects.

In fact, one great triumph of the inflationary scenario was that it provided just such a mechanism. Once a little bit of mass is created by a quantum fluctuation, the inflationary process can pull energy from the gravitational field to make more mass. In this way, the universe is built from mass made by the conversion of gravitational energy.

There are a number of points that usually bother people about this sort of speculation, some easier to clear up than others. For example, the idea that you can drain energy from a gravitational field may be a little disconcerting—where did that energy come from in the first place? Actually, this isn't as severe a problem as it seems, for gravitational energy, unlike some other forms, can be both positive and negative. In other words, it is possible to take energy from a gravitational field, which makes the energy of the field more negative, then take out more energy, making the energy of the field more negative still, and so on. It's somewhat similar to a ball rolling down a long hill—the farther down it rolls the more energy you can get out of it.

A way of visualizing this situation that I find particularly useful is to think about what happens when you dig a hole and pile up the dirt next to it. What you have at the end, of course, is a pile of dirt and a hole. There is no question about where the dirt comes from, or of creating the pile out of nothing. Obviously, the dirt just comes out of the hole. But suppose that for some reason we couldn't see what was happening below ground level. Then the digging process would seem truly miraculous—a pile of dirt would suddenly appear from nowhere. But this wouldn't really be a miracle, because if we took account of everything in the field—dirt plus hole—we would see that the whole process just amounted to a rearrangement, not a creation.

In the same way, when we look at the universe, we tend to notice only the energy tied up in matter (the piles of dirt) and ignore the energy drained from the gravitational field (the holes). With this point of view, it's hardly surprising that we view the creation of matter as miraculous. But as seen by modern physicists, matter is just one more form of energy, a form which can be shifted around at will. Seen in this light, the creation of the universe is no more miraculous than the operation of an ordinary nuclear reactor. Both are just examples of the basic equivalence of matter and energy.

The picture of the origin of the universe that arises from the notions of quantum fluctuations and mass-energy equivalence is readily intelligible. For an undetermined time before the time we label "the beginning," the universe was a vacuum full of evanescent matter. Then, quite by accident, enough fluctuations occurred close enough together to trigger the process by which energy is drained from the gravitational field, and the process of runaway inflation started. This fluctuation, then, was the mixing together of the ingredients of the cosmic "dough." When the period of inflation was over, the Big Bang had begun and, as they say, the rest is history.

The important thing to realize is that in this scheme, no energy is created in the "creation." The total energy of the universe was zero before it started and is still zero today. What happens is that the positive energy locked up in mass is canceled by the negative energy of the gravitational field. The two balance, just as corporate income balances debt in a financial statement.

What I like about this scenario is that it provides a simple, believable, nonmiraculous version of the creation epic. In addition, as a writer I really treasure the kind of statements that people make about it—they provide such juicy quotes. One, by Alan Guth, opens chapter 14. Another, by Edward Tryon, one of the pioneers in the study of quantum fluctuations as an explanation of creation, is one of my favorites. "The universe," he said, "is simply one of those things that happen from time to time."

The Universe in Your Basement

Alan Guth has been investigating the consequences of his inflationary universe and has reached some very interesting results. He started by asking how much matter needed to appear in a fluctuation to start the expansion process off. The answer is surprisingly small—only about twenty-two pounds (10 kg). In other words, the vacuum is so unstable that the appearance of only a few pounds of matter can set off the series of reactions that led to the entire universe.

Having said this, I hasten to add that not just any twenty-two-pound weight will do the job. There is no danger, for example, that a sack of sugar on a supermarket shelf will suddenly trigger the formation of a new universe. The twenty-two pounds have to be compressed into a volume much smaller than that of the proton. It turns out that it's the density that matters, not the total amount of stuff in the fluctuation. When matter/energy is compressed to this extent it is not in a form we would recognize, of course, but in another state of matter that physicists call the "false vacuum." There is no place in the universe today where densities of this type are found, so the universe is, for the moment, stable.

The relatively small amount of energy needed to trigger inflation has led Alan Guth to ask what I consider to be the most provocative and exciting question ever asked in the history of science.

Do the laws of physics allow for the creation of a universe by man-made processes?

Or, to use Guth's more mundane version of the question, could you build a universe in your basement?

With some graduate students, Guth has started to attack this problem. The results of preliminary calculations are illustrated in figure 15.2. If you could create a bit of false vacuum, it would distort the space-time network of our universe, form a kind of neck, and then separate. In effect, it would leave our space and become another universe in its own right.

The fact that it might be possible, even in theory, to initiate such a process is stunning, to say the least. It leads to all sorts

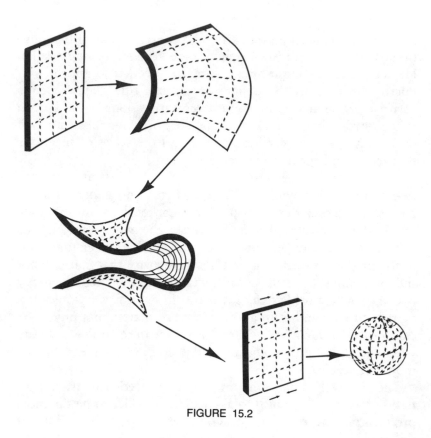

FIGURE 15.2

of incredible possibilities. For example, although creating matter in the state of the false vacuum is not possible today, what if it did become possible at some distant time in the future? Would we then proceed through space-time leaving a trail of expanding universes behind us the way a jet aircraft leaves a condensation trail? Or, more startling still, are *we* the product of a basement experiment in some other, unthinkable universe? Will the human race, having plumbed the secrets of our own universe, now become the creator of others?

This is all speculation, of course—merely dreams. There is little possibility whatsoever that we will be able to build universes in our basements within the lifetime of anyone reading this book. For the moment, then, we have to be content with Alan Guth's summation of our state of knowledge on the origins of the universe. "We cannot yet answer the ultimate questions," he says, "but we can discuss the questions intelligently. We are on the right track."

Epilogue: A Summing Up

WE HAVE FOLLOWED a long path from Isaac Newton's orchard to Alan Guth's speculations about building universes. We have seen how, step-by-step, the domain of the laws of nature have been pushed outward in space and backward in time. At each point in this history there has been a choice. On the one hand, there were those who would have prefered to believe that human knowledge had reached its limits—that henceforeword the world would prove to be different from what it had been thought to be up to that time. On the other, there were those who believed in what we have called the principle of universality. Without exception, the latter prevailed, until now the idea that we can discover here and now the laws that govern the universe everywhere and for all time has become an accepted (and almost unconscious) part of the scientific framework.

It need not have been so. The skies could have remained empty that Christmas Eve in 1758, when Halley's extension of Newton's laws to the comets was triumphantly justified. It could have turned out that Newtonian gravity didn't describe the behavior of the double stars that William Herschel observed from his backyard. The mysterious lines in the solar spectrum that

217

Norman Lockyer identified with terrestrial helium could have remained unidentified. The universe *could* have been much more mysterious than it is. It *could* have turned out that our own little corner of space and time were very different from the other corners—that the universe was a motley collection of variety and discord. It *could* have turned out this way, but it didn't.

In fact, against all reasonable expectation, we have pushed the frontiers of universality to the edges of the observable universe and back to the very beginnings of the cosmos. We have even seen at least one man ask a question that represents the ultimate extension of the principle—Can we build universes ourselves?

In fact, in contemplating this history one can only agree with Albert Einstein when he marveled that "the most incomprehensible thing about the universe is its comprehensibility."

Index